Lecture Notes in Artificial Intelligence 12237

Subseries of Lecture Notes in Computer Science

Series Editors

Randy Goebel
University of Alberta, Edmonton, Canada
Yuzuru Tanaka
Hokkaido University, Sapporo, Japan
Wolfgang Wahlster
DFKI and Saarland University, Saarbrücken, Germany

Founding Editor

Jörg Siekmann
DFKI and Saarland University, Saarbrücken, Germany

More information about this series at http://www.springer.com/series/1244

Wei Lu · Kenny Q. Zhu (Eds.)

Trends and Applications in Knowledge Discovery and Data Mining

PAKDD 2020 Workshops
DSFN, GII, BDM, LDRC and LBD
Singapore, May 11–14, 2020
Revised Selected Papers

 Springer

Editors
Wei Lu (iD)
Singapore University of Technology
and Design
Singapore, Singapore

Kenny Q. Zhu
Shanghai Jiao Tong University
Shanghai, China

ISSN 0302-9743   ISSN 1611-3349 (electronic)
Lecture Notes in Artificial Intelligence
ISBN 978-3-030-60469-1   ISBN 978-3-030-60470-7 (eBook)
https://doi.org/10.1007/978-3-030-60470-7

LNCS Sublibrary: SL7 – Artificial Intelligence

This Springer imprint is published by the registered company Springer Nature Switzerland AG
The registered company address is: Gewerbestrasse 11, 6330 Cham, Switzerland

Preface

The Pacific-Asia Conference on Knowledge Discovery and Data Mining (PAKDD) is one of the longest established and leading international conferences in the areas of data mining and knowledge discovery. It provides an international forum for researchers and industry practitioners to share their new ideas, original research results, and practical development experiences from all KDD-related areas, including data mining, data warehousing, machine learning, artificial intelligence, databases, statistics, knowledge engineering, visualization, decision-making systems, and the emerging applications. PAKDD 2020 was held online, during May 11–14, 2020.

Along with the main conference, PAKDD workshops intend to provide an international forum for researchers to discuss and share research results. After reviewing the workshop proposals, we were able to accept five workshops that covered topics in literature-based discovery, data science for fake news, data representation for clustering, biologically inspired techniques, and game intelligence and informatics. The diversity of topics in these workshops contributed to the main themes of the conference. These workshops were able to accept 17 full papers that were carefully reviewed from 50 submissions. The five workshops were as follows:

- First International Workshop on Literature-Based Discovery (LBD 2020)
- Workshop on Data Science for Fake News (DSFN 2020)
- Learning Data Representation for Clustering (LDRC 2020)
- Ninth Workshop on Biologically Inspired Techniques for Data Mining (BDM 2020)
- First Pacific Asia Workshop on Game Intelligence & Informatics (GII 2020)

May 2020

Wei Lu
Kenny Q. Zhu

Organization

Workshop Co-chairs

Wei Lu   Singapore University of Technology and Design,
                       Singapore
Kenny Q. Zhu   Shanghai Jiao Tong University, China

First International Workshop on Literature-Based Discovery (LBD 2020)

Organizers

Neil R. Smalheiser   University of Illinois at Chicago, USA
Yakub Sebastian   Charles Darwin University, Australia

Workshop on Data Science for Fake News (DSFN 2020)

Organizers

Tanmoy Chakraborty   Indraprastha Institute of Information Technology, India
Deepak P   Queen's University Belfast, UK
Cheng Long   Nanyang Technological University, Singapore
Santhosh Kumar G.   Cochin University of Science and Technology, India
Hridoy Sankar Dutta   Indraprastha Institute of Information Technology, India

Workshop on Learning Data Representation for Clustering (LDRC 2020)

Organizers

Lazhar Labiod   University of Paris, France
Mohamed Nadif   University of Paris, France
Allou Same   l'Institut français des sciences et technologies des
                       transports, de l'aménagement et des réseaux, France

Ninth Workshop on Biologically Inspired Techniques for Data Mining (BDM 2020)

Organizers

Shafiq Alam   The University of Auckland, New Zealand
Gillian Dobbie   The University of Auckland, New Zealand

First Pacific Asia Workshop on Game Intelligence and Informatics (GII 2020)

Organizers

Sharanya Eswaran   Games24x7, India
Magy Seif El-Nasr   Northeastern University, USA
Tridib Mukherjee   Games24x7, India

Contents

Ninth Workshop on Biologically Inspired Techniques for Data Mining (BDM 2020)

First Pacific Asia Workshop on Game Intelligence and Informatics (GII 2020)

First International Workshop on Literature-Based Discovery (LBD 2020)

Matching Biomedical Ontologies with Compact Evolutionary Algorithm

Xingsi Xue[1,2] and Pei-Wei Tsai[3(✉)]

[1] Fujian Key Lab for Automotive Electronics and Electric Drive, Fujian University of Technology, Fujian 350118, China
[2] Intelligent Information Processing Research Center, Fujian University of Technology, Fujian 350118, China
[3] Department of Computer Science and Software Engineering, Swinburne University of Technology, Hawthorn 3122, Australia
ptsai@swin.edu.au

Abstract. Literature Based Discovery (LBD) aims at building bridges between existing literatures and discovering new knowledge from them. Biomedical ontology is such a literature that provides an explicit specification on biomedical knowledge, i.e., the formal specification of the biomedical concepts and data, and the relationships between them. However, since biomedical ontologies are developed and maintained by different communities, the same biomedical information or knowledge could be defined with different terminologies or in different context, which makes the integration of them becomes a challenging problem. Biomedical ontology matching can determine the semantically identical biomedical concepts in different biomedical ontologies, which is regarded as an effective methodology to bridge the semantic gap between two biomedical ontologies. Currently, Evolutionary Algorithm (EA) is emerging as a good methodology for optimizing the ontology alignment. However, EA requires huge memory consumption and long runtime, which make EA-based matcher unable to efficiently match biomedical ontologies. To overcome these problems, in this paper, we define a discrete optimal model for biomedical ontology matching problem, and utilize a compact version of Evolutionary Algorithm (CEA) to solve it. In particular, CEA makes use of a Probability Vector (PV) to represent the population to save the memory consumption, and introduces a local search strategy to improve the algorithm's search performance. The experiment exploits Anatomy track, Large Biomed track and Disease and Phenotype track provided by the Ontology Alignment Evaluation Initiative (OAEI) to test our proposal's performance. The experimental results show that CEA-based approach can effectively reduce the runtime and memory consumption of EA-based matcher, and determine high-quality biomedical ontology alignments.

Keywords: Literature Based Discovery · Biomedical ontology matching · Compact Evolutionary Algorithm

© Springer Nature Switzerland AG 2020
W. Lu and K. Q. Zhu (Eds.): PAKDD 2020 Workshops, LNAI 12237, pp. 3–10, 2020.
https://doi.org/10.1007/978-3-030-60470-7_1

1   Introduction

Literature Based Discovery (LBD) aims at building bridges between existing literatures and discovering new knowledge from them [4]. LBD methods have the potential to revolutionize drug discovery gene-disease association discovery, and many other applications in biomedical domains. Biomedical ontology is such a literature that provides an explicit specification on biomedical knowledge, i.e., the formal specification of the biomedical concepts and data, and the relationships between them. Biomedical ontology can provides the common terminology necessary for biomedical researchers to formally describe their data, enabling better data integration and inter-operability, and therefore facilitating translational discoveries. However, since the biomedical ontologies are developed and maintained by different communities, the same biomedical information or knowledge could be defined with different terminologies or in different context. Biomedical ontology matching can determine the semantically identical biomedical concepts in different biomedical ontologies, which is regarded as an effective methodology to address the biomedical ontologies' heterogeneity problem. Essentially, LBD approaches and biomedical ontology matching methods are of a mutual benefit. Due to the fact that a biomedical ontology often have large-scale concepts and their meaning is semantically complex and ambiguous, it is difficult to efficiently match the biomedical ontologies. Recently, Evolutionary Algorithm (EA) is emerging as a good methodology for optimizing the ontology alignment [14]. Xue et al. respectively propose a hybrid EA [16], cooperative EA [11], interactive EA [12], multi-objective EA [17] and many-objective EA [13] to solve the ontology matching problem. But EA requires huge memory consumption and long runtime, which make EA-based matcher unable to efficiently match biomedical ontologies. In this paper, we use a compact Evolutionary Algorithm (CEA) to optimize biomedical ontology alignments, which makes use of a Probability Vector (PV) to represent the population to save the memory consumption, and introduces a local search strategy to improve the algorithm's search performance.

The rest of the paper is organized as follows: Sect. 2 defines the problem by presenting its optimal model, and describe the similarity measure for calculating two biomedical concepts' similarity value; Sect. 3 presents the CEA-based matcher in details; Sect. 4 presents the experimental configuration and results; finally, Sect. 5 draws the conclusion.

2   Preliminaries

2.1   Biomedical Ontology Matching Problem

We utilize the MatchFmeasure to evaluate a biomedical alignment's quality. For more details on MatchFmeasure, please refer to our work [15]. In the following, the biomedical ontology matching problem is defined:

$$\begin{cases} max & f(X) \\ s.t. & X = (x_1, x_2, \cdots, x_{|C_1|})^T \\ & x_i \in \{1, 2, \cdots, |C_2|\}, i = 1, 2, \cdots, |C_1| \end{cases} \quad (1)$$

where O_1 and O_2 are two biomedical ontologies, $|C_1|$ and $|C_2|$ are respectively the concept number of O_1 and O_2; x_i means the ith pair of concept correspondence, i.e. ith source concept is mapped to target x_ith concept; and the objective function is maximize the alignment's MatchFmeasure.

2.2 Similarity Measure on Biomedical Concept

We utilize a hybrid similarity measure, which takes into consideration the concept's syntax, linguistic and context information, to enhance the confidence of the calculated similarity value. First, for each biomedical concept, the information (the label, comment, and property name) from itself and its context (its direct ascendant and descendants) are put into the corresponding profile. Then, the similarity value is calculated according to the following equation.

$$sim(p_1, p_2) = \frac{\sum_{i=1}^{f} \max_{j=1\cdots g}(sim(p_{1i}, p_{2j})) + \sum_{j=1}^{g} \max_{i=1\cdots f}(sim(p_{1i}, p_{2j}))}{f + g} \quad (2)$$

where p_1 and p_2 are respectively two biomedical concept's corresponding profiles; and $sim(p_{1i}, p_{2j})$ is calculated by integrating the SMOA [7] based syntax measure and Unified Medical Language System (UMLS) [1] based linguistic measure.

$$sim(p_{1i}, p_{2j}) = \begin{cases} 1, & \text{if two element are synonymous in UMLS} \\ SMOA(p_{1i}, p_{2j}), & \text{otherwise} \end{cases} \quad (3)$$

3   Compact Evolutionary Algorithm

In this work, we empirically choose the Gray code, which is a binary encoding mechanism, to encode an alignment. On this basis, we further introduce the Probability Vector (PV) [3] to represent the entire population. Particularly, each element in PV is a probability of being 1 with respect to a solution's corresponding gene bit, i.e. PV's ith element's value is the probability of being 1 with respect to a solution's ith gene bit. Therefore, we can utilize PV to generate various individuals by generating random numbers in [0, 1] and comparing them with PV's elements. In each generation, PV is updated to move to the elite so that the newly generated solutions could be closer to the elite. When the elite keeps unchanged, a local search process is carried out to help CEA jump out of the local optima. In particular, local search process dedicates to improve a solution by searching for a better solution in its neighborhood. In this work, we utilize the binary crossover to implement the local search process, for more details, please see also our work [10]. The pseudo-code of CEA is presented in Algorithm 1.

In the initialization process, we initialize PV by setting all the elements inside as 0.5, and the elite solution by PV (lines 3–6). In the evolving process, CEA utilizes PV to generate a new solution, which simulates the behavior of population-based EA with uniform crossover (line 10), and then execute the mutation on the newly generated solution (lines 11–15). Through the competition between the elite and new solution, we try to update the elite (lines 17–24) and move the PV towards the elite (lines 26–38). When the elite keep unchanged for /$delta$ generations, local search algorithm is executed on the elite (lines 40–47), which tries to help CEA jump out of the local optima.

4 Experimental Results

In order to study CEA-based matcher's performance, the experiment exploits Anatomy track[1], Large Biomed track[2] and Disease and Phenotype track[3] provided by the Ontology Alignment Evaluation Initiative (OAEI)[4]. Table 1 summarizes the main statistics of these tracks.

Table 1. Description on the OAEI's tracks.

Track	Sub-task	Ontology	Scale
Anatomy	MA-HA	Adult Mouse Anatomy (MA)	2,744 classes
		Human Anatomy (HA)	3,304 classes
Large biomed	FMA-NCI	Foundation Model of Anatomy (FMA)	78,989 classes
	FMA-SNOMED	Systemized Nomenclature of Medicine (SNOMED)	122,464 classes
	SNOMED-NCI	National Cancer Institute thesaurus (NCI)	66,724 classes
Disease and phenotype	HP-MP	Human Phenotype Ontology (HP)	33,205 classes
		Mammalian Phenotype Ontology (MP)	32,298 classes
	DOID-ORDO	Human Disease Ontology (DOID)	24,034 classes
		Orphanet Rare Disease Ontology (ORDO)	68,009 classes

We compare the alignment's quality among EA-based matcher [9], CEA-based matcher and OAEI's participants in Tables 2 and 3, we also compare the runtime and memory consumption between EA-based matcher and CEA-based matcher in Table 4. The hardware configurations and the results of OAEI's participants are from OAEI's official website,[5]. CEA's results are the mean values of thirty independent executions. EA's configuration is referred to its literature, and CEA uses the following parameters which represent a trade-off setting obtained in an empirical way to achieve the highest average alignment quality on all tracks.

---

[1] http://oaei.ontologymatching.org/2018/anatomy/index.html.
[2] http://www.cs.ox.ac.uk/isg/projects/SEALS/oaei/2018/.
[3] http://oaei.ontologymatching.org/2018/phenotype/.
[4] http://oaei.ontologymatching.org/2018.
[5] http://oaei.ontologymatching.org/2018/index.html.

Algorithm 1. Compact Evolutionary Algorithm

Input: $maxGen = 3000$, $p_m = 0.05$, $st = 0.1$
Output: ind_{elite}

```
1:  ** Initialization **
2:  δ = 0;
3:  for i = 0; i < PV.length; i++ do
4:      PVᵢ = 0.5;
5:  end for
6:  generate an individual through PV to initialize ind_elite;
7:  ** Evolving Process **
8:  gen = 0;
9:  while gen < maxGen do
10:     generate an individual ind_new through PV;
11:     for i = 0; i < ind_new.length; i++ do
12:         if (rand(0, 1) < p_m then
13:             ind_new,i = (ind_new,i + 1) mod 2;
14:         end if
15:     end for
16:     ** Competition **
17:     [winner, loser] = compete(ind_elite, ind_new);
18:     if winner == ind_elite then
19:         δ = δ + 1;
20:     end if
21:     if winner == ind_new then
22:         ind_elite = ind_new;
23:         δ = 0;
24:     end if
25:     ** Update PV **
26:     for i = 0; i < PV.length; i++ do
27:         if winner[i]==1 then
28:             PV[i] = PV[i] + st;
29:             if PV[i] > 1 then
30:                 PV[i] = 1;
31:             end if
32:         else
33:             PV[i] = PV[i] − st;
34:             if PV[i] < 0 then
35:                 PV[i] = 0;
36:             end if
37:         end if
38:     end for
39:     ** Local Search **
40:     if δ > 30 then
41:         ind_localBest=localSearch(ind_elite);
42:     end if
43:     [winner, loser] = compete(ind_elite, ind_localBest);
44:     if winner == ind_localBest then
45:         ind_elite = ind_localBest;
46:         δ = 0;
47:     end if
48:     gen = gen + 1;
49: end while
50: return ind_elite;
```

- Maximum generation: $maxGen = 3000$;
- Mutation probability: $p_m = 0.05$;
- Neighborhood's scale in local search: $C = 10$;
- Step length for updating PV: $st = 0.1$;

4.1   Comparison on Alignment's Quality and Runtime

We utilize the Friedman's test [2] and Holm's test [5] to carry out the statistical comparison on the alignments' quality. In particular, Friedman's test [2] is first used to figure out whether all the competitors present any difference, and Holm's test [5] is then utilized to determine whether one matcher statistically outperforms others. Under the null-hypothesis, Friedman's test states that all the competitors are equivalent, and if we need to reject it, the computed value $\mathcal{X}_r^2$ must be equal to or greater than the tabled critical chisquare value at the specified level of significance [6]. In this work, we choose a level of significance $\alpha = 0.05$, and we need to consider the critical value for 6 degrees of freedom (since we are comparing 7 matchers), i.e. $\mathcal{X}_{0.05}^2 = 12.592$.

Table 2. Friedman's test on the alignment's quality. Each value represents the f-measure [8], and the number in round parentheses is the corresponding computed rank.

Testing case	AML	LogMap	XMap	DOME	POMAP++	EA	CEA
Anatomy	0.94 (2)	0.89 (5)	0.89 (5)	0.76 (7)	0.89 (5)	0.84 (6)	0.97 (1)
FMA-NCI	0.93 (1.5)	0.92 (3)	0.86 (5.5)	0.86 (5.5)	0.88 (4)	0.78 (7)	0.93 (1.5)
FMA-SNOMED	0.83 (2)	0.79 (4)	0.77 (5)	0.33 (7)	0.40 (8)	0.75 (6)	0.87 (1)
NCI-SNOMED	0.80 (2)	0.77 (3)	0.69 (5)	0.64 (7)	0.68 (6)	0.70 (4)	0.83 (1)
HP-MP	0.84 (3)	0.85 (2)	0.47 (6.5)	0.47 (6.5)	0.68 (5)	0.72 (4)	0.86 (1)
DOID-ORDO	0.64 (6)	0.84 (2)	0.70 (5)	0.60 (7)	0.83 (3)	0.75 (4)	0.89 (1)
Average	0.83 (2.80)	0.84 (3.19)	0.73 (5.50)	0.61 (6.92)	0.72 (5.28)	0.75 (5.35)	0.89 (1.08)

In Table 2, the computed $\mathcal{X}_r^2 = 116.498$, which is greater than 12.592, and therefore, we can reject the null hypothesis and the Holm's test can be further carried out. As shown in Table 2, since CEA-based matcher ranks with the lowest value, it is set as a control matcher that will be compared with others.

Table 3. Holm's test on the alignment's quality.

i	Approach	z–value	Unadjusted p–value	$\frac{\alpha}{k-i}, \alpha = 0.05$
8	AML	2.80	0.043	0.050
7	LogMap	3.19	0.012	0.025
5	POMAP++	5.28	7.42×10^{-7}	0.012
3	EA	5.35	4.90×10^{-7}	0.008
2	XMap	5.50	1.88×10^{-7}	0.007
1	DOME	6.92	5.57×10^{-12}	0.006

In Holm's test, $z-$value is the testing statistic for comparing the ith and jth matchers, which is used for finding the p-value that is the corresponding probability from the table of the normal distribution. p-value is then compared with $\alpha = 0.05$, which is an appropriate level of significance. According to Table 3, it is possible to state that our approach statistically outperforms other biomedical ontology matchers on f-measure at 5% significance level.

Table 4. Comparison on the runtime and memory consumption.

Testing sase	EA runtime (second)	CEA runtime (second)	EA memory (byte)	CEA memory (byte)
Anatomy	88	23	3,016,558	512,108
FMA-NCI	107	32	68,485,120	15,072,236
FMA-SNOMED	172	62	215,591,617	64,219,205
NCI-SNOMED	302	91	72,790,664	12,359,315
HP-MP	166	58	21,328,104	4,254,218
DOID-ORDO	251	73	15,411,731	3,340,209
Average	181	56	66,103,965	16,626,215

Table 4 compares CEA with EA on the runtime and memory consumption. From the Table 4, we can see that CEA can dramatically reduce EA's runtime and memory consumption. To conclude, CEA-based ontology matching technique can effectively reduce the runtime and memory consumption of EA-based matcher, and determine high-quality biomedical ontology alignments.

5   Conclusion

To efficiently optimize the biomedical ontology alignment, in this paper, a discrete optimal model is constructed for the problem, and a CEA-based ontology matching method is presented to solve it. CEA utilizes the compact encoding mechanism to approximate classic EA's evolutionary process, which can dramatically reduce EA's memory consumption. In addition, CEA introduce the local search algorithm, which can further improve the algorithm's performance. The experimental results show that our method can efficiently match various biomedical ontologies.

Acknowledgements. This work is supported by the National Natural Science Foundation of China (No. 61503082), the Natural Science Foundation of Fujian Province (No. 2016J05145), the Program for New Century Excellent Talents in Fujian Province University (No. GY-Z18155), the Program for Outstanding Young Scientific Researcher in Fujian Province University (No. GY-Z160149) and the Scientific Research Foundation of Fujian University of Technology (Nos. GY-Z17162 and GY-Z15007).

References

1. Bodenreider, O.: The unified medical language system (UMLS): integrating biomedical terminology. Nucl. Acids Res. **32**(Suppl_1), D267–D270 (2004)
2. Friedman, M.: The use of ranks to avoid the assumption of normality implicit in the analysis of variance. J. Am. Stat. Assoc. **32**(200), 675–701 (1937)
3. Harik, G.R., Lobo, F.G., Goldberg, D.E.: The compact genetic algorithm. IEEE Trans. Evol. Comput. **3**(4), 287–297 (1999)
4. Henry, S., McInnes, B.T.: Literature based discovery: models, methods, and trends. J. Biomed. Inform. **74**, 20–32 (2017)
5. Holm, S.: A simple sequentially rejective multiple test procedure. Scand. J. Stat., 65–70 (1979)
6. Sheskin, D.J.: Handbook of Parametric and Nonparametric Statistical Procedures. CRC Press, Boca Raton (2003)
7. Stoilos, G., Stamou, G., Kollias, S.: A string metric for ontology alignment. In: Gil, Y., Motta, E., Benjamins, V.R., Musen, M.A. (eds.) ISWC 2005. LNCS, vol. 3729, pp. 624–637. Springer, Heidelberg (2005). https://doi.org/10.1007/11574620_45
8. Van Rijsbergen, C.J.: Foundation of evaluation. J. Doc. **30**(4), 365–373 (1974)
9. Wang, J., Ding, Z., Jiang, C.: GAOM: genetic algorithm based ontology matching. In: 2006 IEEE Asia-Pacific Conference on Services Computing (APSCC 2006), pp. 617–620. IEEE (2006)
10. Xue, X.: Compact memetic algorithm-based process model matching. Soft Comput. **23**(13), 5249–5257 (2018). https://doi.org/10.1007/s00500-018-03672-y
11. Xue, X., Chen, J., Chen, J., Chen, D.: Using compact coevolutionary algorithm for matching biomedical ontologies. Comput. Intell. Neurosci. **2018** (2018)
12. Xue, X., Chen, J., Yao, X.: Efficient user involvement in semiautomatic ontology matching. IEEE Trans. Emerg. Top. Comput. Intell., 1–11 (2018)
13. Xue, X., Lu, J., Chen, J.: Using NSGA-III for optimising biomedical ontology alignment. CAAI Trans. Intell. Technol. **4**(3), 135–141 (2019)
14. Xue, X., Pan, J.S.: An overview on evolutionary algorithm based ontology matching. J. Inf. Hiding Multimed. Signal Process. **9**, 75–88 (2018)
15. Xue, X., Wang, Y.: Optimizing ontology alignments through a memetic algorithm using both MatchFmeasure and unanimous improvement ratio. Artif. Intell. **223**, 65–81 (2015)
16. Xue, X., Wang, Y.: Using memetic algorithm for instance coreference resolution. IEEE Trans. Knowl. Data Eng. **28**(2), 580–591 (2015)
17. Xue, X., Wang, Y.: Improving the efficiency of NSGA-II based ontology aligning technology. Data Knowl. Eng. **108**, 1–14 (2017)

Towards Expert Preference on Academic Article Recommendation Using Bibliometric Networks

Yu Zhang[1(✉)], Min Wang[1], Morteza Saberi[2], and Elizabeth Chang[1]

[1] University of New South Wales, Canberra, ACT 2604, Australia
{yu.zhang,min.wang,e.chang}@adfa.edu.au
[2] University of Technology Sydney, Sydney, NSW 2007, Australia
morteza.saberi@uts.edu.au

Abstract. Expert knowledge can be valuable for academic article rec-ommendation, however, hiring domain experts for this purpose is rather expensive as it is extremely demanding for human to deal with a large vol-ume of academic publications. Therefore, developing an article ranking method which can automatically provide recommendations that are close to expert decisions is needed. Many algorithms have been proposed to rank articles but pursuing quality article recommendations that approx-imate to expert decisions has hardly been considered. In this study, domain expert decisions on recommending quality articles are investi-gated. Specifically, we hire domain experts to mark articles and a com-prehensive correlation analysis is then performed between the ranking results generated by the experts and state-of-the-art automatic ranking algorithms. In addition, we propose a computational model using hetero-geneous bibliometric networks to approximate human expert decisions. The model takes into account paper citations, semantic and network-level similarities amongst papers, authorship, venues, publishing time, and the relationships amongst them to approximate human decision-making fac-tors. Results demonstrate that the proposed model is able to effectively achieve human expert-alike decisions on recommending quality articles.

Keywords: Academic article recommendation · Human expert decision · Bibliometric networks · Data management

1 Introduction

Academic article recommendation has been critical for readers who are seeking for articles with high prestige [12,14], and it would be particularly valuable if the recommendation can be provided by domain experts. However, achieving article recommendation that is approximate to expert decisions has been hardly considered.

Hiring collective human intelligence, known as crowdsourcing, has been fre-quently adopted as an overarching method for large-scale database annotation

W. Lu and K. Q. Zhu (Eds.): PAKDD 2020 Workshops, LNAI 12237, pp. 11–19, 2020.
https://doi.org/10.1007/978-3-030-60470-7_2

and evaluation tasks. It usually requires human input in terms of natural language annotation, computer vision recognition, and answering cognitive questions [2]. The applications of crowdsourcing techniques in these domains have demonstrated its relatively high accuracy and reliability [11].

Although having expert judgement on academic articles would be an ideal way to rank them and recommend the top ones, it is a huge workload for human to manually evaluate each article and rank them accordingly, especially when human resources are rather limited while the workload is heavy. Automatic paper ranking and recommendation system has been a sound solution to handle the enormous and ever growing volume of articles. Classic algorithms include PageRank [5], CoRank [15], FutureRank [7], and P-Rank [10]. In addition, more advanced algorithms, such as HITS [8] and W-Rank [13], have also been proposed recently to integrate more bibliometric information and factors. However, there is a lack of analysis of these algorithms towards human expert decisions. Some studies adopted crowdsourcing approaches to rank only a small amount of top articles and compared the ranking results with those generated by the algorithms for evaluation purposes [9]. Asking crowd workers to investigate the top articles makes sense to certain extent since the essential aim of article ranking is to recommend valuable articles to readers, nonetheless, it remains infeasible for the workers to retrieve the top ones from large-scale article databases. Therefore, it is necessary to design a method which can automatically generate a reasonable article ranking list that approximates to domain expert decisions.

In this study, domain expert decision on ranking academic articles is investigated and a computational model is proposed to approach the expert decisions. Specifically, we analyse the decisions of a group of experts in the research domain of intrusion detection (cyber security) by comparing their ranking results with those generated by state-of-the-art algorithms. Based on the analysis, a computational model is proposed to approximate expert decision on recommending prestige articles using a heterogeneous bibliometric network. The proposed method is fully automatic and is able to generate reasonable ranking results approaching domain expert decisions. The contribution of this study can be summarised into two folds. Firstly, the analysis of domain expert decisions and the comparative study remedy the gap that the existing literature has barely considered to approximate expert decision on recommending quality academic articles. Secondly, the proposed method is able to automatically generate reasonable ranking results close to those from domain experts, breaking through the limitations of human's capability in handling large-scale bibliometric databases.

2   Materials and Methods

2.1   Domain Expert Recommendation Analysis

In this study, three domain experts in the research field of intrusion detection (cybersecurity) are hired, and accordingly, we use bibliometric data in the same field for analysis and evaluation. The data includes 6,428 articles, 16,284 citations, 12,093 authors and 1,048 venues collected from the Microsoft Academic

Graph (MAG) database from year 2000 to 2017. A historical time point is set at year 2008 to divide the database into two parts. The articles published from 2000 to 2008 are used for testing the algorithms, and the remaining articles from 2009 to 2017 are used for evaluating the future trends.

Since it is infeasible for human to deal with such large-scale databases, we use citation count and five popular ranking algorithms, including PageRank, Co-Rank, P-Rank, FutureRank, and HITS as baselines to rank papers in the first partition. Afterwards, only the 30 top articles ranked by each baseline are collected and summarised into a list of 120 articles after removing overlaps, then fed to the experts because recommending top articles is the essential target for academic article recommendation.

Algorithm 1: Converging human expert scores

Input: Score list by each expert: S_i; error rate for each expert: ϵ_i, $i \in \{1, ..., M\}$
Output: Normalised final score list: S

1 $i^* \leftarrow \text{Argmin } \epsilon_i$
2 $Q_1, Q_2 \leftarrow ExhaustiveSearch(i^*, M)$
3 **for** $n = 1 : N$ **do**
4 $\quad d_{ij}(n) \leftarrow |s_i(n) - s_j(n)|; i, j \in \{i^*, Q_1, Q_2\} = M$
5 $\quad$ **if** $size(\text{Argmin } d_{ij}(n)) \leq 1$ **then**
6 $\quad\quad \{i,j\} \leftarrow \text{Argmin } d_{ij}(n)$
7 $\quad\quad S(n) \leftarrow \frac{(1-\epsilon_i)S_i(n)+(1-\epsilon_j)S_j(n)}{2}$
8 $\quad$ **else**
9 $\quad\quad S(n) \leftarrow \frac{\sum_{i \in M}(1-\epsilon_i)S_i(n)}{3}$
10 **return** S

Statistical methods are used to evaluate the expert decisions and converge them into one result. Specifically, the Joglekar's algorithm [3] is used to generate confidence intervals for expert decision error rate estimates by assuming a Bernouli distribution for paper scores made by each expert. Suppose there are M experts and N papers, the error rate of expert i, denoted as ϵ_i, is calculated as follows:

$$\epsilon_i \leftarrow \frac{1}{2} - \sqrt{\frac{\prod_{j \in Q_1, Q_2}(a_{ij} - 1/2)}{2(a_{Q_1 Q_2} - 1/2)}}; \quad M = Q_1 \cup Q_2 \cup \{i\} \qquad (1)$$

where Q_1 and Q_2 are two disjoint sets for expert i, a_{ij} denotes the normalised number of times that expert i and expert j agree on the scores. The agreement of score on paper n between experts i and j is defined as follows:

$$a_{i,j}(n) = \begin{cases} 1 & \text{if } |s_i(n) - s_j(n)| \leq \theta \\ 0 & \text{otherwise;} \quad n \in \{1, ..., N\}; i, j \in \{Q_1, Q_2\} \end{cases} \qquad (2)$$

where θ is a threshold set to 0.05. Exhaustive search strategy is adopted to select the optimal disjoint sets Q_1 and Q_2 which provide the minimum value for $z_t\sqrt{(\epsilon_i(1-\epsilon_i))/N}$, where z_t represents the tth percentile of the standard normal distribution.

Finally, majority voting scheme is used for converging the scores made by multiple experts since it is able to determine a more reasonable score for every article based on the opinions of the majority. The processing procedure is summarised in Algorithm 1.

2.2   Computational Model

Heterogeneous Bibliometric Network. The bibliometric information including articles, authors, venues (journals and conferences), and the relationship amongst them form a heterogeneous network $\mathcal{G}$ which, as illustrated in Fig. 1, can be described with a set of nodes $\mathcal{N}$ and their links $\mathcal{L}$ as follows:

$$\mathcal{G} = \mathcal{G}_{P-A} \cup \mathcal{G}_{P-P} \cup \mathcal{G}_{P-V} \tag{3}$$

$$= \{\mathcal{N}, \mathcal{L}\} = \{\mathcal{N}_A \cup \mathcal{N}_P \cup \mathcal{N}_V, \mathcal{L}_{P-A} \cup \mathcal{L}_{P-P} \cup \mathcal{L}_{P-V}\} \tag{4}$$

where P, A and V denote article, author, and venue, respectively. Considering the citation relevance, the citation network is further updated to $\mathcal{G}_{P-P} = \{\mathcal{N}_P, \mathcal{L}_{P-P}, \mathbf{W}\}$, where $\mathbf{W} \in \mathbb{R}^{N \times N}$ is the adjacency matrix of the citation network and $N = |\mathcal{N}_P|$ is the number of articles in it. The adjacency matrix $\mathbf{W}$ is a representative description of the citation network structure with its entries, denoted as $w_{i,j}$, referring to the relevance of a citation link from article i to article j.

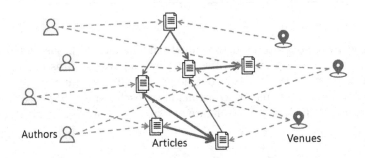

Fig. 1. The heterogeneous bibliometric network

In this study, we follow the definition and implementation of citation relevance in W-Rank [13] which interprets this concept from two perspectives, namely the semantic similarity of the articles' content and the network-level similarity evaluating the mutual links in the citation network. Specifically, we extract the titles and abstracts of articles (as they usually contain the key information of an article), and calculate the semantic similarity of these lexical items using a sense-level algorithm named 'align, disambiguate and walk' [6] which has been demonstrated to be flexible in handling lexical items in different lengths and effective in comparing the meaning of the lexical items. For network-level similarity, we use Cosine similarity to measure the mutual links in the citation network according to the following equation:

$$Cosine(P_i, P_j) = \frac{|L_{P_i} \cap L_{P_j}|}{\sqrt{|L_{P_i}| \times |L_{P_j}|}} \tag{5}$$

where P_i and P_j denote two papers, L_P denotes the links that connect to node P in the citation network, and $L_{P_i} \cap L_{P_j}$ represents the links connecting to both P_i and P_j regardless of the link direction. Finally, the citation relevance is formulated as an integration of the semantic similarity and network-level similarity as follows:

$$w_{i,j} = \alpha \cdot Semantic(P_i, P_j) + \beta \cdot Cosine(P_i, P_j) \tag{6}$$

where α and β are coefficients defined by an exponential function $e^{\lambda(Similarity-\tau)}$, where λ is set to 6 in favour of the similarity values which are greater than the threshold, and the threshold τ is adjusted to be the median values of the two types of similarities, respectively. The α and β are normalised so that $\alpha + \beta = 1$.

Ranking Algorithm. The article ranking algorithm is designed to obtain a score for each article and this is accomplished by propagating between article authority scores S and hub scores H from three types of nodes (paper P, author A, and venue V) in the heterogeneous network.

Specifically, for author-paper network and paper-venue network, the hub score of an author or a venue node X_i is computed according to Eq. 7 and the corresponding authority score propagated from the hub score is computed by Eq. 8.

$$H(X_i) = \frac{\sum_{P_j \in Out(X_i)} S(P_j)}{|Out(X_i)|} \tag{7}$$

$$S_X(P_i) = Z^{-1}(X) \sum_{X_j \in In(P_i)} H(X_j) \tag{8}$$

where $Out(X_i)$ represents the paper nodes linked from the author or venue node X_i in the network, and $In(P_i)$ denotes all the nodes linked to paper node P_i, and $Z(\cdot)$ is a normalisation term.

For citation network, the calculation of the hub score and authority score follows a similar procedure, except that the citation network is a weighted network thus requires the following updates:

$$H(P_i) = \frac{\sum_{P_j \in Out(P_i)} w_{i,j} S(P_j)}{\sum_{P_j \in Out(P_i)} w_{i,j}} \tag{9}$$

$$S_P(P_i) = Z^{-1}(P) \sum_{P_j \in In(P_i)} H(P_j) w_{i,j} \tag{10}$$

where $In(P_i)$ and $Out(P_i)$ denote the nodes linked to and from paper node P_i, respectively.

In addition, we consider publishing time as a factor to promote new papers which are often underestimated by citation-based models due to inadequate citations. An exponential function is used to favour those which are close to the 'Current' time: $S_T(P_i) = Z^{-1}(T)e^{-\rho(T_{Current} - T_{P_i})}$, where $\rho = 0.62$, $T_{Current}$ is the current time of evaluation, and Z is a normalisation term. Finally, the paper authority score S is updated considering the above four components as follows:

$$S(P_i) = \alpha \cdot S_P(P_i) + \beta \cdot S_A(P_i) + \gamma \cdot S_V(P_i) + \delta \cdot S_T(P_i) + (1 - \alpha - \beta - \gamma - \delta) \cdot \frac{1}{N_p} \tag{11}$$

where N_p is the total number of papers in the collection, and the last term represents a random jump. We set the four parameters as $\alpha + \beta + \gamma + \delta + \theta = 0.85$ to allow a 0.15 probability of random jumps. Algorithm 2 summarises the corresponding details.

Algorithm 2: Ranking algorithm

Input: heterogeneous network: $\mathcal{G}_{P-A} \cup \mathcal{G}_{P-P} \cup \mathcal{G}_{P-V}$; publishing time: T_P
Output: paper authority score: S
Parameter: $\alpha, \beta, \gamma, \delta, \tau, \rho$

1 initialise: $S \leftarrow \{1/N_p, 1/N_p, ..., 1/N_p\}$; $old = 1$; $new = -1$;
2 calculate time score: $S_T(P) \leftarrow exp(-\rho(\tau - T_P))$
3 **while** $any(abs(old - new) > 0.0001)$ **do**
4 $\quad$ update hub score and authority score:
5 $\quad$ $H(A) \leftarrow GetHubScore(\mathcal{G}_{P-A}, S)$
6 $\quad$ $H(V) \leftarrow GetHubScore(\mathcal{G}_{P-V}, S)$
7 $\quad$ $H(P) \leftarrow GetHubScore(\mathcal{G}_{P-P}, S)$
8 $\quad$ $S_A(P) \leftarrow GetScore(\mathcal{G}_{P-A}, H(A))$
9 $\quad$ $S_V(P) \leftarrow GetScore(\mathcal{G}_{P-V}, H(V))$
10 $\quad$ $S_P(P) \leftarrow GetScore(\mathcal{G}_{P-P}, H(P))$
11 $\quad$ update paper authority score: $S \leftarrow Integrate(\alpha S_P, \beta S_A, \gamma S_V, \delta S_T, \frac{1}{N_p})$
12 $\quad$ $old = new$; $new = S$;
13 **return** S;

---

3   Results

Based on the database partition, we respectively calculate the citation count (CC), future citation count (FC) and weighted future citations (WFC) generated before and after the historical time point, and then compare them with the domain experts' decisions. Since the objective of this study is to approximate expert decisions on recommending top articles, the ground truth for ranking the articles is the rank list aggregated from the human experts' decisions. Spearman's rank correlation coefficient (ρ) [4] is used to assess the similarity between the ground truth list and the rank list obtained by the selected baselines. We experimentally use the optimal parameters for all the tested methods. The corresponding 0.95 confidence interval (CI) of the estimated ρ is calculated by Fisher transformation [1]: $CI = \tanh(\text{Arctanh}\,\rho \pm z_{\alpha/2}/\sqrt{n-3})$, where N is the sample size and $z_{\alpha/2} = 1.96$ is the two-tailed critical value of the standard normal distribution with $\alpha = 0.05$. In addition, the detection error trade-off (DET, FNR/FPR) curves are generated to compare the classification performance at different thresholds which divide the articles into the highly ranked (positive group) and the lower ranked (negative group).

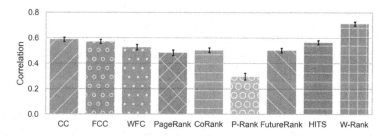

Fig. 2. Spearman's rank correlation between domain expert decision and each method (the error bar indicates a CI of 0.95)

The Spearman's rank correlation results are summarised in Fig. 2. Amongst all the methods, the proposed W-Rank achieved the highest correlation with human expert decisions at 0.714, outperforming the existing algorithms by at least 41.7%. This improvement is also demonstrated by the DET curves shown in Fig. 3. A deeper investigation on the algorithm settings shows that amongst all the bibliometric factors, citation and publication time play a more important role than other factors in approximating ranking results to human expert decisions. However, as the results suggested, it is insufficient to only use citation number to rank articles. Instead, various bibliometric factors should be considered simultaneously, and this is consistent with the decision making process of human experts. In addition, a dramatic drop can be found at the correlation of the P-Rank algorithm in Fig. 2, and the DET curve of the P-Rank in Fig. 3 also covers the largest area under the curve (AUC) indicating poor performance as well. Since P-Rank explored the bibliometric combination of citation, author and venue while all the other baselines either did not use venue factor or added time and other factors, it can be concluded that the venue factor is a redundant factor in approximating expert decisions.

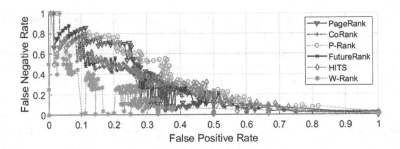

Fig. 3. DET curves of each method

4   Conclusion

Approaching domain expert decisions is a promising direction in designing automatic ranking system for academic article recommendation. In this study, we investigated expert decisions on quality article recommendation by performing a comparative analysis between the results provided by the experts and generated by existing article ranking algorithms. In addition, a computational model was proposed to automatically generate quality article recommendations based on heterogeneous bibliometric networks considering the information of citations, authorship, venues, publishing time, and their relationships. The comparative study and experimental results demonstrate the effectiveness of the proposed model in generating reasonable recommendations that are approximate to expert decisions. It brings an inspiring perspective of integrating both human and machine intelligence into bibliometric ranking and crowdsourcing. In future studies, we will improve the approaches of seeking optimal parameter setting, complementary and redundant features from the bibliometric information for W-Rank using more advanced machine learning methods rather than being experimentally measured.

References

1. Fisher, R.A.: Frequency distribution of the values of the correlation coefficient in samples from an indefinitely large population. Biometrika **10**(4), 507–521 (1915)
2. Jin, Y., Du, L., Zhu, Y., Carman, M.: Leveraging label category relationships in multi-class crowdsourcing. In: Phung, D., Tseng, V., Webb, G., Ho, B., Ganji, M., Rashidi, L. (eds.) PAKDD 2018. LNCS, vol. 10938, pp. 128–140. Springer, Cham (2018). https://doi.org/10.1007/978-3-319-93037-4_11
3. Joglekar, M., Garcia-Molina, H., Parameswaran, A.: Evaluating the crowd with confidence. In: Proceedings of the 19th ACM SIGKDD International Conference on Knowledge Discovery and Data Mining, pp. 686–694. ACM (2013)
4. Myers, J.L., Well, A.D., Lorch Jr., R.F.: Research Design and Statistical Analysis. Routledge, Abingdon (2013)
5. Page, L., Brin, S., Motwani, R., Winograd, T.: The PageRank citation ranking: bringing order to the web. Technical report, Stanford InfoLab (1999)

6. Pilehvar, M.T., Jurgens, D., Navigli, R.: Align, disambiguate and walk: a unified approach for measuring semantic similarity. In: Proceedings of the 51st Annual Meeting of the Association for Computational Linguistics (Volume 1: Long Papers), vol. 1, pp. 1341–1351 (2013)
7. Sayyadi, H., Getoor, L.: FutureRank: ranking scientific articles by predicting their future PageRank. In: Proceedings of the 2009 SIAM International Conference on Data Mining, pp. 533–544. SIAM (2009)
8. Wang, Y., Tong, Y., Zeng, M.: Ranking scientific articles by exploiting citations, authors, journals, and time information. In: Twenty-Seventh AAAI Conference on Artificial Intelligence (2013)
9. Wang, Z., Liu, Y., Yang, J., Zheng, Z., Wu, K.: A personalization-oriented academic literature recommendation method. Data Sci. J. **14** (2015)
10. Yan, E., Ding, Y., Sugimoto, C.R.: P-Rank: an indicator measuring prestige in heterogeneous scholarly networks. J. Am. Soc. Inf. Sci. Technol. **62**(3), 467–477 (2011)
11. Zhang, X., Shi, H., Li, Y., Liang, W.: SPGLAD: a self-paced learning-based crowdsourcing classification model. In: Kang, U., Lim, E.-P., Yu, J.X., Moon, Y.-S. (eds.) PAKDD 2017. LNCS (LNAI), vol. 10526, pp. 189–201. Springer, Cham (2017). https://doi.org/10.1007/978-3-319-67274-8_17
12. Zhang, Y., Saberi, M., Wang, M., Chang, E.: K3S: knowledge-driven solution support system. In: Proceedings of the Twenty-Seventh AAAI Conference on Artificial Intelligence, vol. 33, pp. 9873–9874 (2019)
13. Zhang, Y., Wang, M., Gottwalt, F., Saberi, M., Chang, E.: Ranking scientific articles based on bibliometric networks with a weighting scheme. J. Inf. **13**(2), 616–634 (2019)
14. Zhang, Y., Wang, M., Saberi, M., Chang, E.: From big scholarly data to solution-oriented knowledge repository. Front. Big Data **2**, 38 (2019)
15. Zhou, D., Orshanskiy, S.A., Zha, H., Giles, C.L.: Co-ranking authors and documents in a heterogeneous network. In: Seventh IEEE International Conference on Data Mining (ICDM 2007), pp. 739–744. IEEE (2007)

Research in Progress: Explaining the Moderating Effect of Massively Multiplayer Online (MMO) Games on the Relationship Between Flow and Game Addiction Using Literature-Based Discovery (LBD)

Wendy Hui[1,2]([⊠]) [iD] and Wai Kwong Lau[3] [iD]

[1] Lingnan University, Tuen Mun, Hong Kong
wendyhui@ln.edu.hk
[2] Curtin University, Perth, Australia
[3] University of Western Australia, Perth, Australia
john.lau@uwa.edu.au

Abstract. A recent meta-analytical study reported that there appears to be a weaker correlation between flow and game addiction in players of massively multiplayer online (MMO) games. In this study, we use closed literature-based discovery (LBD) to identify factors that may explain this moderating effect.

Keywords: Literature-based discovery · Massively multiplayer online (MMO) games · Flow · Game addiction · Moderating effect

1 Introduction

Video games have overtaken film, television, music, and books to become the most lucrative sector of the entertainment industry [7]. According to expert estimates, the gaming industry will generate $196 billion in revenue by 2022 [13]. However, as video games continue to gain widespread popularity around the globe, problems arising from game addiction, especially among the young, have begun to attract public attention. In response, governments and game developers have introduced various measures to minimize the social effects of game addiction. For example, in 2011 the South Korean government passed the Shutdown Law, which prevents anyone under 16 from playing games online between midnight and 6 a.m. [7]. In 2019, Tencent, the Internet giant in China, launched a new "anti-addiction system" to restrict the playing time for players aged 18 years and under. The company also introduced a "parental care" platform that allows parents to connect to their children's game accounts and remotely block their children from playing [16].

To better inform policymakers, it is important to understand the factors that can affect game addiction. Li et al. [8] conducted a meta-analysis of empirical game addiction studies that focused on the influence of "flow," which is a term that Csikzentmihalyi

© Springer Nature Switzerland AG 2020
W. Lu and K. Q. Zhu (Eds.): PAKDD 2020 Workshops, LNAI 12237, pp. 20–29, 2020.
https://doi.org/10.1007/978-3-030-60470-7_3

[3] coined to describe the mental state of being fully immersed in an activity. Although flow is often associated with positive experiences, Csikszentmihalyi [4] suggested that addiction is a possible negative effect of flow. According to the brain-disease model of addiction [2], addiction is characterized by the desensitization of the brain's reward system and increased responses to the substance or activity the individual is dependent upon. Taken together, these studies suggest that the experience of flow in games is more likely to trigger abnormal responses in the brain's reward system. Li et al. [8] confirmed a significant positive correlation between flow and addiction in general. Furthermore, they observed that the reported correlations between flow and addiction tend to be smaller in studies that focus on massively multiplayer online (MMO) games. This moderating effect was validated using multiple meta-analytic techniques.

In this study, our objective was to use closed literature-based discovery (LBD) to explore factors that can potentially explain the moderating effect of MMO games on the relationship between flow and game addiction. To the best of our knowledge, this is one of the first studies in which LBD is used to explain a moderating factor. In the following sections, we summarize the findings of Li et al. [8], describe our research methodology, report our findings, and identify factors that can potentially explain the moderating effect. Finally, we conclude the paper and describe our future research directions.

2 The Moderating Effect of MMO

Li et al. [8] conducted a comprehensive search of multiple databases and found 23 empirical studies on game addiction that reported relevant statistics for their meta-analysis. The 23 studies are listed in the Appendix. All of the included studies either directly reported the correlation between game addiction (the dependent variable) and "immersion" or "flow" (the independent variable), or provided sufficient information for us compute the correlation. All of the studies measured the independent variable ("immersion" or "flow" or "escapism") using a comparable measurement scale developed based on [15]. Similarly, all of the studies measured the dependent variable ("addiction" or "problematic play") using a comparable measurement developed based on the Diagnostic and Statistical Manual of Mental Disorders (DSM-IV) [1] and the Internet Addiction Test (IAT) [17].

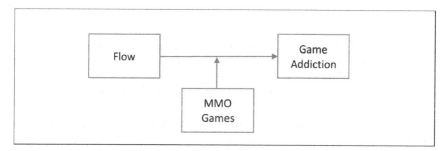

Fig. 1. The moderating effect of MMO games.

We categorized the studies based on the research context into (1) MMO games and (2) non-MMO games. The Q test for homogeneity, credibility intervals, and the Schmidt and Hunter 75% rule were used to detect the moderation effect, which was confirmed by a meta-regression and subgroup analysis. We concluded that although there is an overall positive correlation between flow and addiction, the correlation is weaker in MMO games, as depicted in Fig. 1.

3   Research Methodology

LBD can be either open or closed [14]. As we aimed to identify factors that may affect a specific relationship (between flow and game addiction), our use of closed LBD was appropriate. Using Swanson's [11] ABC model, our "A" concept was flow and our "C" concept was game addiction. Our task was to identify possible paths between A and C for a set of studies on MMO games and for another set of studies on non-MMO games. We wished to identify the path differences in MMO games and non-MMO games that may provide possible explanations of the different correlations observed in the two game types.

Hence, we conducted four different searches, two for each game type, as depicted in Fig. 2. Note that the strengths of the links between A and C for the same B terms may be different in the two game-types. Moreover, the list of B terms linking A and C may also be different.

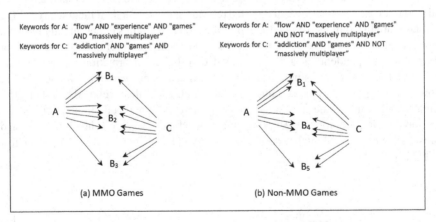

Fig. 2. Search strategy for MMO games and non-MMO games.

In this paper we report some of our preliminary findings. We performed a search on the ProQuest database on January 24, 2020. We limited our search to research articles published in scholarly journals. We generated four sets of documents by using the following keyword searches of abstracts only:

1. Flow (A) in MMO studies: "flow" AND "experience" AND "games" AND "massively multiplayer;"

2. Addiction (C) in MMO studies: "addiction" AND "games" AND "massively multiplayer;"
3. Flow (A) in Non-MMO studies: "flow" AND "experience" AND "games" AND NOT "massively multiplayer;" and
4. Addiction (C) in Non-MMO studies: "addiction" AND "games" AND NOT "massively multiplayer."

The terms "experience" and "games" were added to the search to ensure that only the relevant abstracts were retrieved. Unlike previous LBD studies such as [5] and [12], we did not require the relationship between concept A and concept C to be a novel one. Hence, in the "flow" sets, we did not exclude the term "addiction." Likewise, in the "addiction" sets, we did not exclude the term "flow". This is because we wished to explore the potential moderators of a known relationship rather than explain a novel relationship. Since studies investigating the relationship between flow and addiction may also explore other factors that moderate the relationship, excluding these studies would be too restrictive for our research purpose.

As this search resulted in too few studies in the "Flow (A) in MMO studies" category (only five non-duplicated studies), on March 3, 2020, we conducted another search on the Elsevier Engineering Village and IEEE Xplore using the same set of keywords. Table 1 shows the total number of non-duplicated abstracts retrieved for each set in the combined search from ProQuest, Elsevier Engineering Village, and IEEE Xplore.

Table 1. Number of abstracts in each document set.

Document set	Number of documents
Flow (A) in MMO studies	9
Addiction (C) in MMO studies	59
Flow (A) in Non-MMO studies	740
Addiction (C) in Non-MMO studies	925

As shown, the number of MMO studies mentioning "flow" remained relatively small. In the future, we will need to collect more data using other databases.

4 Data Analysis

To analyze the text data, we used the tm and SnowballC packages in R. For each set of documents, we first transformed all of the text to lower case, and then removed all numbers, English stopwords, and extra white spaces, and finally stemmed all the words. The stopwords were simply the original list of stopwords from the tm package and are presented in the Appendix. For each set of documents, we created a term-document matrix for the unigrams and bigrams. Common terms were identified between the "flow" and "addiction" sets for both MMO games and Non-MMO games. These terms comprised the potential B terms linking the concepts of flow and addiction.

We obtained four lists of B terms, as follows: (1) common unigrams for the MMO games, (2) common bigrams for the MMO games, (3) common unigrams for the non-MMO games, and (4) common bigrams for the non-MMO games. The lengths of the lists were 264, 119, 4470, and 8256, respectively. The B lists for the non-MMO games were much longer than those for MMO games, because there were more non-MMO documents in our dataset.

Lexical statistics were used to evaluate the B terms [9]. Specifically, for each B term in each list, the average tf-idf (term frequency–inverse document frequency) was computed for both the "flow" and the "addiction" sets. The B terms in each list were then ranked by the sum of the average tf-idf of the "flow" set and the average tf-idf of the "addiction" set. From each list, we eliminated words that were too general (e.g., "games," "internet," "player") or irrelevant (e.g., "questionnaire," "research," "study focus"). This step was mostly based on the authors' subjective judgment. Finally, we removed terms that were related to "addiction" and "flow" because we were only interested in potential moderators. After eliminating the irrelevant terms, the top 15 common B terms from each list were identified, as shown in Table 2.

Table 2. Common B terms.

Common unigrams in MMO studies	Common bigrams in MMO studies[a]	Common unigrams in non-MMO studies	Common bigrams in non-MMO studies
Student	Recreation specialization	Students	Social network
Special	Dynamic progression	Learning	Mobile phone
Motives	Inexperience players	Social	Mental health
Social	Personal motives	Gambling	College student
Network	Player interaction	Design	Time spent
Progression		Mobile	School student
Recreation		Education	Social media
Skills		Active	Public health
Interact		Interaction	Young people
Reward		Motives	University student
Performance		Children	Mean age
Entertain		School	Middle school
Dynamic		Media	Prefrontal cortex
Inexperience		Virtual	Mobile device
Leisure		Cognitive	Mental disorder

[a] As there are few MMO studies in our data set, fewer than 15 bigrams are short-listed for the MMO studies.

5   Discussion

Among the common B terms listed in Table 2, the use of medical terms such as "prefrontal cortex," "mental health," and "mental disorder" in the non-MMO literature suggests that these studies recognize game addiction as a health or mental problem. Relevant factors that may link the concept of flow to addiction include age, education, game platform (mobile), and social networks. In contrast, the common B terms in the MMO studies tend to refer to game design features such as "recreation specialization" and "dynamic progression." The relevant factors that may affect the link between flow and addiction include experience, reward, performance and skills. It is interesting to note that "motives" appear in both the MMO and non-MMO literature. Similarly, player interaction and social interaction are similar concepts and appear in both the MMO and non-MMO literature.

Among the common B terms, we believe that age, personal motives, and social interaction can potentially moderate the relationship between flow and addiction. However, factors such as age and personal motives are likely to be relevant to both MMO and non-MMO games, whereas social interaction is more likely to be essential to MMO game experience. Hence, we believe that social interaction is more likely to explain the moderating effect observed in Li et al. [8]. As our data set only includes a small number of MMO studies and our interpretation is subjective, further literature reviews and theory development involving domain experts are needed to validate our proposition.

6   Conclusion and Future Research

In this study, we aimed to explain the moderating effect of MMO games identified in the exploratory meta-analysis conducted by Li et al. [8]. Based on Swanson's [11] ABC model, we conducted an LBD study of articles collected from ProQuest, Elsevier Engineering Village, and IEEE Xplore using "flow" as concept A and "addiction" as concept C. We divided our search into two groups, MMO studies and non-MMO studies, to identify different B terms relating to the two game types and to increase the likelihood of identifying factors that may explain the weakened relationship between flow and addiction in MMO games. Our subjective interpretation of the LBD results identified social interaction as a possible explanation.

A major limitation of our paper is that few MMO studies were found to have contained the keyword "flow." In the future, we will include more databases in our search and involve domain experts in our interpretation of the common B terms. Finally, we will analyze our data using other well established LBD software such as Arrowsmith [12], BITOLA [6], and LION LBD [10].

Appendix

List of Stopwords

i	me	my	myself	we	our
ours	ourselves	you	your	yours	yourself
yourselves	he	him	his	himself	she
her	hers	herself	it	its	itself
they	them	their	theirs	themselves	what
which	who	whom	this	that	these
those	am	is	are	was	were
be	been	being	have	has	had
having	do	does	did	doing	would
should	could	ought	i'm	you're	he's
she's	it's	we're	they're	i've	you've
we've	they've	i'd	you'd	he'd	she'd
we'd	they'd	i'll	you'll	he'll	she'll
we'll	they'll	isn't	aren't	wasn't	weren't
hasn't	haven't	hadn't	doesn't	don't	didn't
won't	wouldn't	shan't	shouldn't	can't	cannot
couldn't	mustn't	let's	that's	who's	what's
here's	there's	when's	where's	why's	how's
a	an	the	and	but	if
or	because	as	until	while	of
at	by	for	with	about	against
between	into	through	during	before	after
above	below	to	from	up	down
in	out	on	off	over	under
again	further	then	once	here	there
when	where	why	how	all	any
both	each	few	more	most	other
some	such	no	nor	not	only
own	same	so	than	too	very

List of Studies Included in [8]

1. Ballabio, M., Griffiths, M. D., Urbán, R., Quartiroli, A., Demetrovics, Z., Király, O.: Do gaming motives mediate between psychiatric symptoms and problematic

gaming? An empirical survey study. Addiction Research & Theory 25(5), 397-408 (2017).

2. Chou, T. J., Ting, C. C.: The role of flow experience in cyber-game addiction. CyberPsychology & Behavior 6(6), 663-675, (2003).

3. Dauriat, F. Z., Zermatten, A., Billieux, J., Thorens, G., Bondolfi, G., Zullino, D., Khazaal, Y.: Motivations to play specifically predict excessive involvement in massively multiplayer online role-playing games: Evidence from an online survey. European Addiction Research 17(4), 185-189, (2011).

4. Hagström, D., & Kaldo, V.: Escapism among players of MMORPGs: Conceptual clarification, its relation to mental health factors, and development of a new measure. Cyberpsychology, Behavior, and Social Networking 17(1), 19-25, (2014).

5. Hull, D. C., Williams, G. A., & Griffiths, M. D.: Video game characteristics, happiness and flow as predictors of addiction among video game players: A pilot study. Journal of behavioral addictions 2(3), 145-152, (2013).

6. Jung, D., Kim, E. Y., Jeong, S. H., & Hahm, B. J.: Higher immersive tendency in male university students with excessive online gaming. International Symposium on Pervasive Computing Paradigms for Mental Health, 157-161. Springer (2015).

7. Kardefelt-Winther, D.: Problematizing excessive online gaming and its psychological predictors, Computers in Human Behavior 31, 118-122 (2014).

8. Khan, A., Muqtadir, R.: Motives of problematic and nonproblematic online gaming among adolescents and young adults. Pakistan Journal of Psychological Research 31(1), 119. (2016).

9. Khang, H., Kim, J. K., Kim, Y.: Self-traits and motivations as antecedents of digital media flow and addiction: the Internet, mobile phones, and video games. Computers in Human Behavior 29(6), 2416-2424 (2013).

10. Kuss, D., Louws, J., Wiers, R. W.: Online gaming addiction? Motives predict addictive play behavior in massively multiplayer online role-playing games. CyberPsychology, Behavior, and Social Networking 15(9), 480-485, (2012).

11. Kwok, N. W. K., Khoo, A.: Gamers' motivations and problematic gaming: An exploratory study of gamers in World of Warcraft. International Journal of Cyber Behavior, Psychology, and Learning (IJCBPL) 1(3), 34-49 (2011).

12. Li, D. D., Liau, A. K., Gentile, D. A., Khoo, A., Cheong, W. D.: Construct and predictive validity of a brief MMO player motivation scale: Cross-sectional and longitudinal evidence based on Singaporean young gamers. Journal of Children and Media, 7(3), 287-306, (2013).

13. Liu, C. C., Chang, I. C.: Model of online game addiction: The role of computer-mediated communication motives. Telematics and Informatics 33(4), 904-915, (2016).

14. Oggins, J., Sammis, J.: Notions of video game addiction and their relation to self-reported addiction among players of World of Warcraft. International Journal of Mental Health and Addiction 10(2), 210-230, (2012).

15. Park S., Hwang H.S.: Understanding online game addiction: connection between presence and flow. In Jacko J.A. (ed) Human-Computer Interaction: Interacting in Various Application Domains. Lecture Notes in Computer Science 5613. Springer, Berlin, Heidelberg (2009).

16. Seah, M. L., Cairns, P.: From immersion to addiction in videogames. Proceedings of the 22nd British HCI Group Annual Conference on People and Computers: Culture, Creativity, Interaction 1, 55-63. British Computer Society (2008).

17. Snodgrass, J. G., Dengah, H. F., Lacy, M. G., Fagan, J.: A formal anthropological view of motivation models of problematic MMO play: achievement, social, and immersion factors in the context of culture. Transcultural Psychiatry 50(2), 235-262 (2013).

18. Snodgrass, J. G., Lacy, M. G., Dengah, H. F., Eisenhauer, S., Batchelder, G., Cookson, R. J.: A vacation from your mind: problematic online gaming is a stress response. Computers in Human Behavior 38, 248-260 (2014).

19. Sun, Y., Zhao, Y., Jia, S. Q., Zheng, D. Y.: Understanding the antecedents of mobile game addiction: the roles of perceived visibility, perceived enjoyment and flow. Pacfic Asia Conference on Information Systems, 141 (2015).

20. Wan, C. S., Chiou, W. B.: Psychological motives and online games addiction: A test of flow theory and humanistic needs theory for Taiwanese adolescents. CyberPsychology & Behavior 9(3), 317-324 (2006).

21. Wu, T.-C., Scott, D., Yang, C.-C.: Advanced or addicted? Exploring the relationship of recreation specialization to flow experiences and online game addiction. Leisure Sciences 35(3), 203-217 (2013).

22. Xu, Z., Turel, O., Yuan, Y.: Online Game addiction among adolescents: Motivation and prevention factors. European Journal of Information Systems 21(3), 321-340 (2012).

23. Zhong, Z., Yao, M. Z.: Gaming motivations, avatar-self-identification and symptoms of online game addiction. Asian Journal of Communication 23(5), 555-573 (2013).

References

1. American Psychiatric Association: Diagnostic and Statistical Manual of Mental Disorders (DSM-4 ed.). American Psychiatric Pub. (1994)

2. Butler Centre for Research: the brain disease model of addiction. (2016). https://www.hazeld enbettyford.org/~/media/files/bcrupdates/bcr_ru02_braindiseasemodel.pdf?la=en. Accessed Jan 29 2019

3. Csikszentmihalyi, M.: Beyond Boredom and Anxiety. Jossey-Bass Publishers, San Francisco (1975)

4. Csikszentmihalyi, M.: Flow: The Psychology of Happiness. Jossey-Bass Publishers, San Francisco (1992)

5. Gordon, M.D., Lindsay, R.K., Fan, W.: Literature-based discovery on the world wide web. ACM Trans. Internet Technol. 2(4), 261–275 (2002)

6. Hristovski, D., Peterlin, B., Mitchell, J.A., Humphrey, S.M.: Using literature-based discovery to identify disease candidate genes. Int. J. Med. Inform. 74, 289–298 (2005)

7. Jabr, F.: Can you really be addicted to video games? N. Y. Times (2019)

8. Li, M., Hui, W., Reiners, T., Wongthongtham, P.: Break the flow: a meta-analysis of the relationship between flow and game addiction in massively multiplayer online games. Working paper, Curtin University (2020)

9. Lindsay, R.K., Gordon, M.D.: Literature-based discovery by lexical statistics. J. Am. Soc. Inf. Sci. **50**(7), 574–587 (1999)

10. Pyysalo, S., et al.: LION LBD: a literature-based discovery system for cancer biology. Bioinformatics **35**(9), 1553–1561 (2019)

11. Swanson, D.R.: Fish oil, Raynaud's syndrome, and undiscovered public knowledge. Perspect. Biol. Med. **30**(1), 7–18 (1986)

12. Swanson, D.R., Smalheiser, N.R.: An interactive system for finding complementary literatures: a stimulus to scientific discovery. Artif. Intell. **91**(2), 183–203 (1997)

13. Web, K.: The $120 billion gaming industry is going through more change than it ever has before, and everyone is trying to cash in. BusinessInsider.com, October 1 2019

14. Weeber, M., Kors, J.A., Mons, B.: Online tools to support literature-based discovery in life sciences. Brief. Bioinform. **6**(3), 277–286 (2005)

15. Yee, N.: Motivations for play in online games. CyberPsychology Behav. **9**(6), 772–775 (2006)

16. Ying, X.: Who's to blame for game addiction? News China (2019)

17. Young, K.S.: Caught in the Net: How to Recognize the Signs of Internet Addiction-and a Winning Strategy for Recovery. Wiley, Hoboken (1998)

A Network Approach for Mapping and Classifying Shared Terminologies Between Disparate Literatures in the Social Sciences

Cristian Mejia$^{(\boxtimes)}$ and Yuya Kajikawa

Tokyo Institute of Technology, Tokyo 108-0023, Japan
`mejia.c.aa@m.titech.ac.jp`

Abstract. Methodologies for literature-based discovery in their closed modality, aim to connect two disjoint literatures A and C by providing a list of B-terms from where hypotheses can be drawn. In this paper, we propose a method for representing B-terms as a network. Such representation further allows us to obtain shared themes and ease the exploration of common terms. This is done by firstly dividing the A and C literatures into narrower topics and extracting intersecting lists of B-terms between those. The lists are then used to compute a term cooccurrence network. The method can be applied to fields where curated or standardized thesaurus does not exist, like in the social sciences. We illustrate the method by linking the disparate literatures on poverty alleviation and the Internet of Things. One being the first and most pressing social issue targeted by the Sustainable Development Goals and the other a rapidly growing concept that encompasses technological solutions. We expect our method contributes as an exploratory tool to navigate literature-based discovery outputs easily.

Keywords: Literature-based discovery · Citation networks · Text mining · Poverty alleviation · Internet of Things

1 Introduction

Literature-based discovery refers to a collection of text mining methodologies whose purpose is to connect disjoint literatures. In doing so, new hypotheses can be created systematically, thus potentially accelerating the discovery of solutions to known problems [1]. For instance, in his seminal work Swanson [2] established a connection between Raynaud's syndrome and fish oil as potential treatment, by mining academic articles on both topics and finding linking terms as "blood viscosity" to be shared in common. The treatment was later validated trough clinical trials.

Despite text mining methods being well spread across most fields of science, literature-based discovery methods are still focused on biomedical research [3]. Fewer studies are found in the social sciences, mainly focused in discovery for innovation, policy, and technology roadmap creation [4]. Even fewer are the works exploring the use of citation networks as means of potential discovery.

© Springer Nature Switzerland AG 2020
W. Lu and K. Q. Zhu (Eds.): PAKDD 2020 Workshops, LNAI 12237, pp. 30–40, 2020.
https://doi.org/10.1007/978-3-030-60470-7_4

Among those, we find the work of Ittipanuvat et al. [5] who applied citation networks to link the topics of gerontology and robotics, and Fujita [6] who utilized a similar approach to establish connections between Sustainability and complex networks. In both cases, the researchers successfully identified patterns of relationship between their A and C literatures at the expense of seemingly laborious inspection and validation by experts. In the case of Ittipanuvat et al., a similarity matrix in the form of a heatmap was provided and the pairs of clusters deemed to be similar were retrieved. Under such approach it is required a close inspection of the articles in each pair of clusters for the expert to notice what exactly is the logical connection between them. On the other hand, Fujita also provided a list of B-terms for each connection.

Our proposed method differs from those in that we reach a greater topical resolution and provide an output that clearly defines the themes between the two literatures on the study. Concretely, the objective of this paper is to develop a method for the graphical representations of B-terms that ease the exploration of common themes and concepts between literatures A and C.

To illustrate the method, we connect the literatures of poverty reduction and the Internet of Things (IoT). The first corresponds to the first goal of the Sustainable Development Goals (SDG) "end poverty in all its forms everywhere" [7] and the second is an umbrella term collecting several technologies expected to enhance quality of life and exert great economic impact [8]. Then, poverty reduction literature represents a collection of social issues while research on IoT provides a collection of solutions. We expect that our method helps to start shedding light on the common themes between both research areas where synergies could be possible.

2 Data and Methods

As in the classic ABC approach of literature-based discovery we start by obtaining the sets of articles related to the concepts of interest A and C, here "poverty alleviation" and "IoT" respectively. We retrieved bibliographic data from the Web of Science Core Collection, which covers articles from the sciences, social sciences, arts and humanities. A topical search (TS) was performed to find any document matching the queries shown in Table 1 in either the title, abstract or keywords.

Setting a query for identifying research related to the SDGs have proven not to be a trivial task, and experimental studies have been made to set the combination of terms that retrieves the most related documents per each goal [9]. Hence, we used verbatim the suggested query from that study. On the other hand, given that IoT is a coined term, we retrieved articles by setting its full and abbreviated writing as the query. Data were retrieved on January 10th, 2020.

We take advantage of the analysis of gaps which is the linking of subfields of literatures A and C [10]. In this case those subfields are automatically identified by following the method described in Fig. 1. We created a citation network where each article is represented as a node connected to each other node they cite. Among the different modalities of citation networks, direct citation networks as described here are known to bring better topical representations [11]. Because we are interested in articles directly related to the concept A or C, we retain the largest connected component of each network and neglect

Table 1. Selected queries for research on poverty alleviation and smart cities.

Query	Articles
TS = (("extreme poverty" OR "poverty alleviation" OR "poverty eradication" OR "poverty reduction" OR "international poverty line" OR ("financial aid" AND "poverty") OR ("financial aid" AND "poor") OR ("financial aid" AND "north-south divide") OR ("financial development" AND "poverty") OR "financial empowerment" OR "distributional effect" OR "distributional effects" OR "child labor" OR "child labour" OR "development aid" OR "social protection" OR "social protection system" OR ("social protection" AND access) OR microfinanc* OR micro-financ* OR "resilience of the poor" OR ("safety net" AND "poor") OR ("economic resource" AND access) OR ("economic resources" AND access) OR "food bank" OR "food banks")); Timespan: All years	21,409
TS = ("internet of things" OR "iot"); Timespan: All years	45,728

the disconnected ones. Also, we intend to bridge two disjoint literatures. Hence, articles intersecting both large components are removed and inspected separately if necessary. These intersecting articles are removed from the large components of A and C so that they get completely disjoint. Then we proceeded to split the networks into clusters.

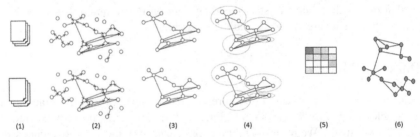

(1)   (2)   (3)   (4)   (5)   (6)

Fig. 1. Methodological approach: records for literatures A and C are retrieved (1). citation networks are built (2) and the largest connected component is selected (3). Both networks are clustered (4), the content similarity of each pair of clusters is computed and the B-terms extracted for each intersection (5). A network of b-term co-occurrences is built (6).

Clusters were identified by applying a modularity maximization algorithm. The modularity Q measures how well divided is a network into clusters by comparing the strength of intra-cluster connections versus the inter-cluster connections. Is defined as follows:

$$Q = \sum_i \left(e_{ii} - a_i^2\right), a_i = \sum_j e_{ij} \tag{1}$$

Where e_{ij} is the fraction of edges connecting *cluster i* and *cluster j*. While e_{ii} is the fraction of edges within *cluster i*. The maximization of Q and the selection of the number of clusters was computed with the Louvain algorithm [12], which scales well for large networks and is broadly used to cluster citation networks in particular.

At this point, we have the optimal clusters based on the modularity value for both networks representing the major clusters for each of them. We labeled the clusters after inspecting the contents of their most cited articles. Thus, obtaining an academic land-scape of both literatures. We then proceeded to obtain subclusters by dividing each major cluster using the same methodological approach.

The purpose of using subclusters is twofold. First, they allow us to split the literatures into fine-grained clusters and avoid the problem of resolution limit [13] found in large networks. As their size gets reduced, they become more cohesive and easier to interpret. To be more specific, their vocabulary is also expected to be narrower. Secondly, the more subclusters found, the more pairwise comparisons are possible between subclusters of both networks. This number of intersections will be an artifact that helps to create the network of co-occurring B-terms.

To find the intersections, we compute the content similarity of subclusters between the networks. First, text content needs to be preprocessed and cleaned. We concatenated the title, abstract, and keywords of each article, removed stop words, lowercased and steamed them. Then we proceeded to compute the *tfidf* weights for the terms in the clusters as follows:

$$w_c^{(i)} = tf_{i,c} \cdot \log\left(\frac{N}{df_i}\right) \tag{2}$$

$tf_{i,c}$ is the frequency of term i in subcluster c, df_i is the number of documents where i appears, and N is the total number of documents. w_c is normalized, so that $\|w_c\| = 1$. These vectors are used to compute the following similarity scores:

$$Cosine(c_1, c_2) = w_{c_1} \cdot w_{c_2} \tag{3}$$

$$Jaccard(c_1, c_2) = \frac{w_{c_1} \cdot w_{c_2}}{\sum_i w_{c_1}^2 + \sum_i w_{c_2}^2 - w_{c_1} \cdot w_{c_2}} \tag{4}$$

$$Dice(c_1, c_2) = \frac{2(w_{c_1} \cdot w_{c_2})}{\sum_i w_{c_1} + \sum_i w_{c_2}} \tag{5}$$

$$Simpson(c_1, c_2) = \frac{|w_{c_1} \cdot w_{c_2}|}{\min(|w_{c_1}|, |w_{c_2}|)} \tag{6}$$

Where w_c is the *tfidf* vector computed with (2). Similarity scores are calculated for each pair of subclusters between networks. We selected the pairs having an above-average similarity score to extract their intersecting terms, or B-terms. We evaluate the robustness of the method by obtaining the correlations between similarity scores. A high correlation among the scores signals that the same pairs of clusters were picked as highly similar and ranked likewise. Thus, bridging themes would be consistent regardless of the similarity metric. Finally, a B-terms cooccurrence network is built, and clusters of terms are found by applying the Louvain algorithm.

3   Results

We found that research on poverty alleviation can be divided into 18 major clusters covering the 11,561 articles in the largest component. Articles span from the '70 s with an outlier article from 1922 on child labor [14]. On the other hand, IoT research is divided into 16 clusters containing the 28,099 articles of the largest connected component. IoT is a relatively younger field of research with the earliest paper using the concept being published in 1993. The lists of clusters are summarized in Table 2 in Table 3 and respectively.

Table 2. Major clusters in the citation network of poverty alleviation

Cluster	Cluster label	Articles	Avg. Year	Avg. Citations
1	Agriculture and economic growth	1374	2012.6	19.7
2	Microfinance and access to credit	1314	2013.6	12.4
3	Child labor	1166	2009.8	15.5
4	Social transfer/social protection	1158	2013.3	15.3
5	Biodiversity conservation	997	2012.7	32.2
6	Microfinance institutions	958	2014.4	12.3
7	Rural poverty	552	2011.1	22.4
8	Access to Energy	478	2014.2	16.2
9	Aid	413	2013.0	13.6
10	Entrepreneurship and CSR	384	2014.4	20.7
11	World Bank and development goals	374	2011.9	15.8
12	HIV and poverty	335	2014.0	23.0
13	Food security	326	2014.8	18.0
14	Public health systems	325	2013.8	23.3
15	Tourism and poverty reduction	285	2013.1	15.5
16	Welfare	284	2012.5	21.3
17	Carbon tax	202	2011.6	20.7
18	Others	636	2010.2	18.1

We also found six articles simultaneously having both queries. However, none of them were connected to the main literature corpus of either network represented by the largest connected component, thus not considered in the study. Four of these articles appeared given incidental mentions of keywords related to poverty or IoT without being the central topic of the articles. The remaining 2 explore the topic of IoT for agriculture with attention to rural development [15, 16].

The earliest research on poverty alleviation focused on child labor and rural poverty, while entrepreneurship and corporate social responsibility (CSR), and food security are the newer trends based on the average publication year of their articles. The cluster with more attention in the academic community based on the average citations received is biodiversity conservation.

Table 3. Major clusters in the citation network of IoT

Cluster	Cluster label	Articles	Avg. Year	Avg. Citations
1	Smart manufacturing	3499	2017.0	10.2
2	Security and privacy	3082	2017.4	6.5
3	Low-power wide-area network	2599	2017.7	7.8
4	Edge computing	2579	2017.9	7.9
5	Social IoT	2562	2016.5	8.3
6	Policy and academic discussions on IoT	1915	2016.0	7.7
7	Smart cities	1729	2017.1	7.1
8	Sensors	1380	2016.9	7.0
9	IoT optimization frameworks	1367	2017.5	7.3
10	Middleware	1209	2017.1	5.6
11	Protocols and architectures	937	2017.1	5.8
12	Ambient backscatter	931	2017.6	7.5
13	Blockchain and smart contracts	916	2017.9	7.5
14	Smart grid integration	686	2017.4	5.9
15	Bluetooth	597	2017.2	5.8
16	Others	2111	2017.4	8.2

The difference between the oldest and youngest cluster on IoT research is near to 2 years, with the overall publication year average being 2017.2. Therefore, difficult to discriminate between mature or emergent trends. The topic of smart manufacturing is the one receiving more attention in academic research, both by the number of publications and the average citations received.

Those major clusters were further divided into subclusters. We found a total of 179 subclusters for poverty alleviation and 267 for IoT research. Then, we computed the content similarity of all possible pairs of subclusters between the two networks and retained those pairs above average. Table 4 shows the correlation between the scores of pairs regarded as highly similar using four similarity scores. We found an exact correlation between cosine and dice similarity, followed by Jaccard. Simpson's similarity also correlates to the others. Therefore, it is observed that consistent results can be obtained regardless of the similarity score for the present study. Figure 2 shows the cooccurrence network derived from the complete list of B-terms from pairs of subclusters having an above-average cosine similarity score.

In the network, each B-term is connected to another if they cooccurred pair of subclusters. The size of the nodes represents the degree or number of connections and the thickness of the edges the number of times the two terms cooccur. Terms that appeared two times or more are shown. We found 8 clusters of B-terms: innovation, agriculture, knowledge creation, technology adoption, supply chain, corporate social responsibility,

Table 4. Correlation among four similarity scores.

	Cosine	Jaccard	Dice	Simpson
Cosine	1.0000			
Jaccard	0.9999	1.0000		
Dice	1.0000	0.9999	1.0000	
Simpson	0.7837	0.7823	0.7837	1.0000

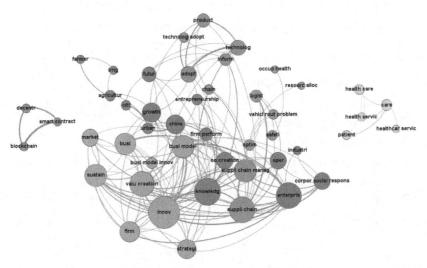

Fig. 2. Cooccurrence network of B-terms

healthcare, and blockchain, which can be said to be the themes bridging the literatures of poverty alleviation and IoT.

Finally, we compare our results with other two common methods used in literature-based discovery. The first is the classic ABC approach, as explained by Swanson and Smalheiser [17]. The second is a particular implementation of the aforementioned approach called Arrowsmith [18], which is an easy-to-use web interface that extracts B-terms based on articles from the Medline database. Table 5 shows the top B-terms output by these methods. For the first method we focused on 2-words keywords only, given that unigrams were too general. For Arrowsmith, we applied the same queries in Table 1; however, being integrated with Medline, it is expected that the output is biased towards biomedicine keywords.

Table 5. B-terms obtained using two classic literature-based discovery methods

Method	Top 15 B-terms
Swanson and Smalheiser (1997)	Case study; health care; energy efficiency; big data; business model; supply chain; social network; resource allocation; developing country; climate change; decision making; energy consumption; systematic review; renewable energy; performance evaluation
Arrowsmith	Antiretroviral therapy; antiretroviral; electronic health; carbon nanotube; highly active antiretroviral; ionic liquid; electronic health record; active antiretroviral therapy; big data; decision support; cell phone; social network; sustainable; HIV testing; mobile phone

4   Discussion and Conclusion

In this article, we have proposed a methodological approach to map B-terms as a network, thus enabling their classification by finding clusters of terms tightly connected. This is expected to help experts in reducing the burden on exploring such lists and have an overview of the *themes* that connect both literatures.

We linked the literatures on poverty alleviation to the one of IoT. Both are umbrella terms covering a variety of topics from where connections could be established. It is worth noting that at the current state, despite the large volume of publications in both sets, the direct connections between them are limited, with only two articles on IoT for agricultural applications in the intersection [15, 16]. Both articles did not have references to the main corpus of literature under study, thus seen as peripheral cases.

Being the research on poverty alleviation and IoT disjoint, only indirect connections are possible. Our proposed method helped to surface such linkage, finding eight transversal themes between them. These are represented as a network of B-terms.

The network of B-terms is composed of general and specific terms. Hence, knowledge domain is also advised to correctly navigate the output. For instance, keywords as "product," "technology," and "future" are general and tend to pull several articles in literatures A and C. However, as the keywords are grouped into cluster is easier to find concepts that might bring interesting connections between the literatures and ease the discovery process. As a practical example, we focus on 3 of the clusters we deemed to be more specific. First, the case of "supply chain." The literature on supply chain on IoT research can be found scattered across subclusters of smart manufacturing. We found the development of solutions for animal tracking and health checking using inexpensive RFID tags, wireless sensor networks, and 3D printing technologies feasible to apply for rural areas in the bovine livestock production cycle [19]. Hence, suitable for developing countries. This IoT subcluster is found to share content similarity to the Entrepreneurship cluster of the poverty alleviation network.

Another intersecting theme is healthcare. Major cluster 14 in the network of poverty alleviation research tackles on this topic covering public health systems, including discussions on making health more accessible and affordable. As a counterpart, IoT research explores developments in decision support systems integrated in Health-IoT devices that help reduce cost in diagnosis [20]. Monitoring devices for pregnant women and the introduction of the concept of community healthcare system, a network of monitoring devices in rural areas is also present [21]. Research that is aligned with the intention of cost reduction for healthcare.

The cluster composed of the B-terms blockchain, smart contract, and decentralization points to the application of blockchain, or distributed ledger technologies to transfer and manage assets in the context of the poverty alleviation loan program in China [22]. This kind of application is incipient in the context of poverty alleviation, but its underlying technology can be enhanced by the many developments linking blockchain and IoT, where at least 155 articles discuss this intersecting topic.

A comparison of the proposed method to other known literature-based discovery approaches found similarities. Keywords, like health care, business model, supply chain, were also observed when applying the classic ABC approach. But other related to energy and climate change were missing in our network. On the other hand, those related to agriculture, innovation, and corporate social responsibility were surfaced by the network. Still, more work is needed to understand in which scenario a method should be favored over the other. Still, the network approach attempts to offer easier navigation to potentially long lists of B-terms, thus expecting to speed-up the discovery process.

By linking poverty alleviation to IoT research by following the proposed approach, a policymaker working on an IoT project would focus her or his attention on strengthening features of the project that are related to any of the clusters found in the map. This would make the project be in line with the known narratives found in research for poverty alleviation. Moreover, the connection of the terms to the specific subclusters helps to narrow down the literature from where ideas can be drawn for the project to address the goal of poverty reduction. Additionally, the map could help for the completeness of the project by finding those themes or terms not yet covered by the project. Hence, the map does not establish a direct linkage but helps to rethink IoT policies and technology roadmaps to address pressing social issues like poverty reduction is.

The clear limitation of this method is the lack of evaluation. While we evaluated the potential consistency of the results regardless of the similarity metric applied, further study is needed to evaluate both the quality of the clusters obtained and the impact on the discoveries made. However, the lack of a gold-standard or accepted evaluative framework is not specific to the present study, but rather a common problem to be solved throughout literature-based discovery research [3, 23]. Another limitation is that the method was illustrated using a single case. More comparisons, for instance, with other SDGs or policy concepts, might help to sum up evidence on the usability of literature-based discovery beyond biomedical research.

Notwithstanding, we expect our method contributes to the literature-based discovery research as an exploratory tool and as a visual aid that helps bring sense to the shared terms between target literatures. Future efforts will be directed towards the evaluation and improvements of such output.

References

1. Kostoff, R.N.: Systematic acceleration of radical discovery and innovation in science and technology. Technol. Forecast. Soc. Change **73**, 923–936 (2006). https://doi.org/10.1016/j.techfore.2005.09.004
2. Swanson, D.: Fish oil, Raynaud's syndrome, and undicovered public knowledge. Perspect. Biol. Med. **30**, 7–18 (1986)
3. Sebastian, Y., Siew, E.-G., Orimaye, S.O.: Emerging approaches in literature-based discovery: techniques and performance review. Knowl. Eng. Rev. **32**, e12 (2017). https://doi.org/10.1017/S0269888917000042
4. Kostoff, R.N., Boylan, R., Simons, G.R.: Disruptive technology roadmaps. Technol. Forecast. Soc. Change **71**, 141–159 (2004)
5. Ittipanuvat, V., Fujita, K., Sakata, I., Kajikawa, Y.: Finding linkage between technology and social issue: a literature based discovery approach. J. Eng. Technol. Manag. JET-M **32**, 160–184 (2014). https://doi.org/10.1016/j.jengtecman.2013.05.006
6. Fujita, K.: Finding linkage between sustainability science and technologies based on citation network analysis. In: Proceedings of the 5th IEEE International Conference on Service-Oriented Computing and Applications, SOCA 2012 (2012)
7. United Nations: Transforming our world: the 2030 Agenda for Sustainable Development (2015)
8. Unlocking the potential of the Internet of Things | McKinsey. https://www.mckinsey.com/business-functions/mckinsey-digital/our-insights/the-internet-of-things-the-value-of-digitizing-the-physical-world. Accessed 30 Jan 2020
9. Jayabalasingham, B.; Boverhof, R.; Agnew, K.; Klein, L.: Identifying research supporting the United Nations sustainable development goals, vol. 1. Mendeley (2019)
10. Smalheiser, N.R.: Rediscovering Don Swanson: the past, present and future of literature-based discovery. J. Data Inf. Sci. **2**, 43–64 (2018). https://doi.org/10.1515/jdis-2017-0019
11. Klavans, R., Boyack, K.W.: Which type of citation analysis generates the most accurate taxonomy of scientific and technical knowledge? J. Assoc. Inf. Sci. Technol. **68**, 984–998 (2017). https://doi.org/10.1002/asi.23734
12. Blondel, V.D., Guillaume, J.-L., Lambiotte, R., Lefebvre, E.: Fast unfolding of communities in large networks. J. Stat. Mech: Theory Exp. **2008**, P10008 (2008). https://doi.org/10.1088/1742-5468/2008/10/P10008
13. Fortunato, S., Barthélemy, M.: Resolution limit in community detection. Proc. Natl. Acad. Sci. U. S. A. **104**, 36–41 (2007). https://doi.org/10.1073/pnas.0605965104
14. Fuller, R.G.: Child labor and child nature. Pedagog. Semin. **29**, 44–63 (1922). https://doi.org/10.1080/08919402.1922.10534004
15. Qu, D., Wang, X., Kang, C., Liu, Y.: Promoting agricultural and rural modernization through application of information and communication technologies in China. Int. J. Agric. Biol. Eng. **11**, 1–4 (2018). https://doi.org/10.25165/IJABE.V11I6.4428
16. Dlodlo, N., Kalezhi, J.: The internet of things in agriculture for sustainable rural development. In: Proceedings of 2015 International Conference on Emerging Trends in Networks and Computer Communications, ETNCC 2015 (2015)
17. Swanson, D.R., Smalheiser, N.R.: An interactive system for finding complementary literatures: a stimulus to scientific discovery. Artif. Intell. **91**, 183–203 (1997). https://doi.org/10.1016/S0004-3702(97)00008-8
18. Smalheiser, N.R., Torvik, V.I., Zhou, W.: Arrowsmith two-node search interface: A tutorial on finding meaningful links between two disparate sets of articles in MEDLINE. Comput. Methods Programs Biomed. **94**, 190–197 (2009). https://doi.org/10.1016/j.cmpb.2008.12.006

19. Addo-Tenkorang, R., Gwangwava, N., Ogunmuyiwa, E.N., Ude, A.U.: Advanced animal track-&-trace supply-chain conceptual framework: an internet of things approach. Procedia Manufacturing **30**, 56–63 (2019)
20. Yang, Y., et al.: GAN-based semi-supervised learning approach for clinical decision support in health-IoT platform. IEEE Access **7**, 8048–8057 (2019). https://doi.org/10.1109/ACCESS.2018.2888816
21. Nazir, S., Ali, Y., Ullah, N., García-Magariño, I.: Internet of Things for healthcare using effects of mobile computing: a systematic literature review. Wirel. Commun. Mob. Comput. (2019)
22. Wang, H., Guo, C., Cheng, S.: LoC — a new financial loan management system based on smart contracts. Futur. Gener. Comput. Syst. **100**, 648–655 (2019). https://doi.org/10.1016/j.future.2019.05.040
23. Smalheiser, N.R.: Literature-based discovery: beyond the ABCs. J. Am. Soc. Inf. Sci. Technol. **63**, 218–224 (2012). https://doi.org/10.1002/asi.21599

Towards Creating a New Triple Store for Literature-Based Discovery

Anna Koroleva[1,2](✉) , Maria Anisimova[1,2], and Manuel Gil[1,2](✉)

[1] Institute of Applied Simulation, School of Life Sciences and Facility Management,
Zurich University of Applied Sciences (ZHAW),
Grüentalstrasse 14, P.O. Box, 8820 Waedenswil, Switzerland
aakorolyova@gmail.com, manuel.gil.sci@gmail.com
[2] Swiss Institute of Bioinformatics (SIB), Quartier Sorge - Bâtiment Génopode,
1015 Lausanne, Switzerland

Abstract. Literature-based discovery (LBD) is a field of research aiming at discovering new knowledge by mining scientific literature. Knowledge bases are commonly used by LBD systems. SemMedDB, created with the use of SemRep information extraction system, is the most frequently used database in LBD. However, new applications of LBD are emerging that go beyond the scope of SemMedDB. In this work, we propose some new discovery patterns that lie in the domain of Natural Products and that are not covered by the existing databases and tools. Our goal thus is to create a new, extended knowledge base, addressing limitations of SemMedDB. Our proposed contribution is three-fold: 1) we add types of entities and relations that are of interest for LBD but are not covered by SemMedDB; 2) we plan to leverage full texts of scientific publications, instead of titles and abstracts only; 3) we envisage using the RDF model for our database, in accordance with Semantic Web standards. To create a new database, we plan to build a distantly supervised entity and relation extraction system, employing a neural networks/deep learning architecture. We describe the methods and tools we plan to employ.

Keywords: Literature-based discovery · Triple store · Semantic web · Information extraction

1 Introduction

Literature-based discovery (LBD) is a research field aiming at discovering new knowledge by mining scientific literature. The field was pioneered in 1980s by Don Swanson, who discovered previously unknown connections between Raynaud's disease and dietary fish oil [46] and between migraine and magnesium [47]. Swanson formulated the first LBD paradigm, called the ABC paradigm: term A and term C may occur in non-intersecting sets of publications (i.e. A and C

This work is funded by a grant (9710.3.01.5.0001.08) from Health@N, ZHAW.

W. Lu and K. Q. Zhu (Eds.): PAKDD 2020 Workshops, LNAI 12237, pp. 41–50, 2020.
https://doi.org/10.1007/978-3-030-60470-7_5

never occur in the same article), but they can be related via some intermediate terms B; discovering the link between A and C via B is a classical task of LBD.

As noted in a thorough literature review of Gopalakrishnan et al. [17], many of the existing LBD systems work with pre-processed input, often using existing information extraction systems or knowledge bases storing the information extracted from the publications (entities, relations, etc.). One of the largest and most commonly used databases is SemMedDB [24], created with the use of SemRep [42], a rule-based system for extracting semantic relations. The great value of SemRep and SemMedDB for LBD is demonstrated by the number of works using them [8,9,13,19,41,43,50,54]. However, SemRep and SemMedDB have some limitations:

- SemRep relies on MetaMap [2] for extracting entities. MetaMap is a rule-based system for mapping texts to the UMLS (Unified Medical Language System [5]) thesaurus. As any lexical resource, UMLS has the drawback consisting in the delay between the appearance of new entities in the literature and their introduction in the thesaurus, which can be problematic for LBD research concerning emerging investigational agents. As a consequence, MetaMap has limited coverage with regard to certain entity types [31,53]. Despite its wide use, MetaMap is outperformed by statistical methods [1]. Last but not least, MetaMap (and its lighter version MetaMap Lite) is slow which makes it difficult to use in large-scale text processing: Neumann et al. [35] compared the speed of a few biomedical NLP pipelines and showed that MetaMap Lite is one of the slowest systems, taking 89 ms per sentence, while e.g. the Python scispaCy library, based on neural networks, takes 4 ms/sentence.
- SemRep extracts only 30 predicate types, hence it does not cover all the possible relations. Even within the addressed predicate types, some verbs with relevant semantics are missed by the system, e.g. SemRep extracts the predications from test sentences "Dietary fish oil **increases/improves** blood pressure", but not from the test sentences "Dietary fish oil **decreases/lowers** blood pressure". SemRep mainly focuses on verb predicates and does not extract many important predicative phrases containing adjectives or nouns, e.g. the relations in the sentences such as "Platelet aggregation **is high in** patients with Raynaud's disease" and "Dietary fish oil **is associated to/causes an improvement in** blood viscosity" are not extracted.
- SemRep employs rules to extract relations. Although this method is well-established and tends to show good precision, it has recently been outperformed by machine learning methods [7].

Our goal is to create a new database of relations extracted from biomedical publications, extending SemMedDB by adding more types of entities and predicates, and using recent deep learning methods. We plan to represent the relations as triples (subject - predicate - object), in accordance with Semantic Web standards, accompanied by a number of meta-data parameters. In order to create such a database, we plan to employ Natural Language Processing (NLP) to extract entities and relations from texts of publications. In the following sections, we outline several directions of our work and describe NLP methods and tools that we plan to employ.

2 Our Contribution: An Extended Knowledge Base

2.1 New Types of Predicates, Entities and Relations

As we stated above, SemRep only covers 30 predication types and misses some of the predicates relevant to these types. Our first goal thus is to add the relevant predicates missed by SemRep (e.g. verb "to_lower", "be_high", "cause_improvement").

LBD arose in the medical domain, and the majority of existing use cases of LBD concern relations between diseases and potential drugs. However, new applications emerge, both in the medical domain where LBD is used to discover interactions involving genes, proteins, cells, receptors, biological processes, etc. (e.g. [3, 10, 12, 20, 23, 25, 37, 45, 48, 49], and in other domains (see [22]). Hence, our second goal is to extract the types of entities and relations (for instance provided by biomedical ontologies) not targeted by SemRep/SemMedDB. In particular, in consultation with an expert in Natural Product Chemistry and Phytopharmacy, we identified the following potential applications for LBD (entities are in italics, predicates are in bold):

1. Control of due diligence in terms of ABS (access and benefit sharing) regulations. Relations to be extracted:
   - *Product X* **contains** *the natural product Y*.
   - *Natural product Y* **is contained in** *the biological species N*.
   - *The biological species N* **is native in** *the countries B, C, D*.
   - *Countries B and C* **have** *ABS regulations* in place.

   Discovery: *Product X* **falls under ABS regulations of** *countries B, C*.
2. "How green is your medicine?". Relations to be extracted:
   - *Plant X* **is sourced from** *country Y*.
   - *Plant X* **is endangered** with status ZZ in *country Y*.
   - *Plant X* **is contained in** *(phyto)pharmaceuticals QQ*.

   Discovery: *QQ* **are not green**.
3. Use of natural products in traditional medicine. Relations to be extracted:
   - *Plant X* **is traditionally used as a cure for disease B**.
   - *Plant X* **contains** *compound Y*.
   - *Compound Y* **is active against** *disease Z*.

   Discovery: *Plant X* **might be active against** *disease Z*.

The new types of entities of interest include: location (country), biological species, regulatory document names. The new types of relations include: "contain", "be_native_in", "be_sourced_from", "be_endangered_in", etc.

2.2   Data Sources

All the current LBD systems that we are aware of process titles and abstracts only. This approach is justified by a number of reasons, such as the full text being unavailable, available at a high cost and in formats inconvenient for automatic processing [21], as well as scalability issues. Still, using full texts has a potential benefit for LBD [21].

In our work, we plan to leverage full texts of PubMed Open Access dataset, as well as MEDLINE abstracts. The drawback of this decision is the high work-load and, consecutively, scalability issues. However, the full texts provide more complete information about the findings of a paper than its abstract and title. Besides, the Introduction section of full text articles normally contains refer-ences to other works and their findings, which can provide an additional source of information: the cited papers might not be available themselves, hence the citations are the only source for extracting such information. We believe that the benefits of using the full texts outweigh the drawbacks —a hypothesis we will be able to test with our system.

Other potentially useful sources of information include: preprints (available at arXiv[1], bioRxiv[2], medRxiv[3]), conference proceedings (abstracts and full papers), presentations and posters, internet sources such as DrugBank[4] (already used by some works [11]). We are not planning to employ these resources for the moment, but they present a promising source of information for future LBD research.

2.3   Database Design

We plan to implement our data base using Semantic Web technologies in accor-dance with the standards set by the World Wide Web Consortium[5]. Therefore, the knowledge base will be deployed as a triple store using the Resource Descrip-tion Framework (RDF) data model. RDF stores can readily be accessed via the RDF query language SPARQL.

Further, the biomedical and bioinformatics communities use various domain specific ontologies for annotation and database integration. They provide a com-mon vocabulary, machine interpretable definitions of concepts, and relations among them. Notable ontologies in the context of this work are ChEBI (Chemical Entities of Biological Interest [18]), NCBITaxon (representing the NCBI organ-ismal taxonomy [16]), Open PHACTS (Open Pharmacological Concepts [51]), UMLS, or the pharmacokinetics ontology [52]. Such ontologies are included in the OBO Foundry [44], a collaborative effort, which includes about 140 ontologies and is now considered a gold standard. Specifically, the OBO Foundry has estab-lished principles for ontology development and evolution, with the aim of maxi-mizing cross-ontology coordination and interoperability. OBO can be obtained in

---

[1] https://arxiv.org/.

[2] https://www.biorxiv.org/.

[3] https://www.medrxiv.org.

[4] https://www.drugbank.ca/.

[5] https://www.w3.org/standards/semanticweb/.

RDF format. We will use a selection of relevant ontologies for entity recognition and to enrich the knowledge base with additional meaning.

Certain metadata items in a database of relations can be useful for LBD research. We plan to include the following metadata: the part of the paper where a given relation occurs (title, abstract, Introduction, Results/Conclusions); the count of occurrences of relations in the corpora; the modality of the relation (negative, affirmative, uncertain).

3   Methods

3.1   Overview of the Approach

SemRep employs simple dictionary look-up and rules. These methods are widely used as they can show good precision and do not require annotated data. However, for complex tasks such as biomedical entity and relation extraction, these methods are generally outperformed by machine learning (ML) methods. ML methods based on neural networks and deep learning recently established the new state of the art for many NLP tasks, including entity and relation extraction (e.g. [14,30,39,40]). Hence, we plan to use neural networks/deep learning architecture as the basis of our approach.

Supervised learning using high-quality manually annotated data shows the best performance across different ML approaches. However, annotated data are sparse, there are no currently available corpora for certain types of relations that interest us. Annotating data for all possible types of relations is effort-intensive and time-consuming. Hence, we plan to employ distantly supervised learning.

Our approach is inspired by the framework for unsupervised relation extraction proposed by Papanikolaou et al. [38], which in turn partially follows the OpenIE framework [4]. Their approach consists in the following:

1. the types of entities of interest are to be pre-defined; entities are extracted using an existing algorithm;
2. the list of predicates of interest is defined;
3. syntactic parsing is used to extract the shortest dependency path between pairs of entities occurring in the same sentence. If there are any verbs ($V1$, $V2$, etc.) in the dependency path, the pair of entities is considered to be connected via relations V1, V2, etc. Word embeddings are created for the verbs using a pre-trained skipgram model [32]. Cosine similarity between the extracted verbs and the defined set of predicates is calculated; if the similarity between a verb Vx and a predicate Py is above a set threshold, the pair of entities is tagged as having the relationship of the type Py. This procedure is used to automatically annotate a corpus, to be used for distantly supervised relation extraction.
4. the automatically annotated corpus is used to fine-tune BERT as a relation extraction classifier.

The approach was tested on four datasets (the Biocreative chemical-disease relations dataset [29], the Genetic Association Database [6], the EUADR dataset [34], and a proprietary dataset Healx CD) and was slightly inferior to supervised models trained for each dataset, but superior to other unsupervised approaches.

We implement a similar approach with a few modifications, presented below.

3.2 Creation of a Corpus for Distant Supervision

To extract entities, we plan to employ the Python library spaCy[6] and its modification for scientific language processing scispaCy [35]. SpaCy provides functionality for extracting a wide range of general-domain entities, including locations and titles of laws. ScispaCy provides several models trained on various biomedical corpora for extracting a wide range of biomedical entities[7]. SpaCy and scispaCy use convolutional neural network models for entity extraction.

An important biomedical entity not extracted by spaCy/scispaCy is outcomes of clinical trials. To extract outcomes, we will use the algorithms that we developed in our previous research [26]. For other types of entities not extracted by spaCy/scispaCy, we will seek to find an existing tool for their extraction or an existing corpus that we can train on (e.g. the COPIOUS corpus [36] for species names, MedMentions [33] for a wide range of UMLS entities).

For relation extraction, we are considering multi-word predicates, where the main semantics is not necessarily expressed by the verb (e.g. "be high", "be associated with an increase"). Thus, we plan to calculate the similarity of the pre-defined predicates with the whole sequence of words between the entities. We tested a few algorithms for assessing semantic similarity, including gensim similarity, the bert-embeddings package and our previously developed model [27]. The majority of algorithms fails to adequately assess similarity of phrases that substantially differ in length. Our model [27], which is a fine-tuned BioBERT [28] model, handles this problem well, according the results of a small test set.

3.3 Distantly Supervised Relation Extraction

We plan to fine-tune BioBERT for relation extraction. Alternatively, we consider employing a joint entity and relation extraction algorithm such as SPERT [15] to train the final model on the distantly annotated dataset.

4 Conclusion and Difficulties

We plan to create an extended database for LBD. We envisage several challenges and difficulties. Scalability of the approach is the biggest issue. To tackle it, we need to filter the sentences before applying the relation extraction at a large scale. The best approach to filtering sentences is under question.

---

[6] https://spacy.io/.

[7] See https://allenai.github.io/scispacy/.

The second issue is limited coverage of existing entity extraction systems. The majority of relation extraction approaches consider only sentences with (at least) two extracted entities. However, in certain cases, one entity in a relation of interest might not be extracted by the algorithms used. This may be due to errors of entity extraction algorithms or some entity types not being covered by the algorithms employed. For example, in the sentence "Dietary fish oil lowers blood viscosity." the entity "Dietary fish oil" is not extracted by any of spaCy/scispaCy models. We plan to train additional entity extraction algorithms for entity types missed by their models; however, this still does not guarantee a full coverage of all the entity types of interest. One possible way of dealing with this problem is to process all sentences with at least one entity extracted and consider all the noun phrases (NPs) that occur within a certain distance to be a potential second entity. In this case, we would assess the similarity of the pre-defined predicates and the context between the entity and each NP. If the context and a predicate are similar, we would consider the entity and the NP to be related via the corresponding predicate. Determining the type of the new entity (the NP) remains an issue.

Another difficulty consists in developing a consistent scheme for representing the extracted relations. The relations are not always easily represented as "entity1 - predicate - entity2". Consider the examples:

1. A increases B.
2. A is associated with an increase in B.
3. A is associated with increased B.
4. A is associated with B.

Examples 2 and 3 need to be represented in the same way, different from both 1 and 4. Still, in the future LBD research using the database, a researcher interested in the example 1 or example 4 is likely to also want to find the examples 2 and 3. Hence, we need to develop a scheme to consistently describe all the variations in patterns.

Finally, we expect that full texts contain a higher proportion of sentences with coreference relations, compared to abstracts. Coreference relations need to be interpreted correctly in order to extract meaningful triples. Thus we need to include a coreference resolution algorithm in our pipeline.

We will address the listed challenges in our future work. We plan to release the new database, as well as the developed information extraction modules. We expect the extensions proposed here to improve on the state of the art of LBD. It will be an essential part of our work to evaluate if this expectation is met.

References

1. Abacha, A.B., Zweigenbaum, P.: Medical entity recognition: a comparison of semantic and statistical methods. In: BioNLP ACL (2011)
2. Aronson, A.: Effective mapping of biomedical text to the UMLS Metathesaurus: the metamap program. In: AMIA Annual Symposium 2001, pp. 17–21, February 2001

3. Baker, N.C.: Methods in literature-based drug discovery (2010)
4. Banko, M., Cafarella, M.J., Soderland, S., Broadhead, M., Etzioni, O.: Open information extraction from the web. In: Proceedings of the 20th International Joint Conference on Artificial Intelligence, IJCAI 2007, Morgan Kaufmann Publishers Inc., San Francisco, CA, USA, pp. 2670–2676 (2007)
5. Bodenreider, O.: The unified medical language system (UMLs): integrating biomedical terminology. Nucleic Acids Res. **32**(Database issue), D267–D270 (2004)
6. Bravo, A., Piñero, J., Queralt-Rosinach, N., Rautschka, M., Furlong, L.I.: Extraction of relations between genes and diseases from text and large-scale data analysis: implications for translational research. BMC Bioinform. **16**, 55 (2015)
7. Bui, Q.C.: Relation extraction methods for biomedical literature. Ph.D. thesis, Informatics Institute (IVI), University of Amsterdam (2012)
8. Cairelli, M.J., Miller, C.M., Fiszman, M., Workman, T.E., Rindflesch, T.C.: Semantic MEDLINE for discovery browsing: using semantic predications and the literature-based discovery paradigm to elucidate a mechanism for the obesity paradox. In: AMIA Annual Symposium Proceedings, pp. 164–73 (2013)
9. Cameron, D., Kavuluru, R., Rindflesch, T.C., Sheth, A.P., Thirunarayan, K., Bodenreider, O.: Context-driven automatic subgraph creation for literature-based discovery. J. Biomed. Inf. **54**, 141–157 (2015)
10. Chen, H., Sharp, B.M.: Content-rich biological network constructed by mining PubMed abstracts. BMC Bioinform. **5**, 147 (2004)
11. Chichester, C., Digles, D., Siebes, R., Loizou, A., Groth, P., Harland, L.: Drug discovery FAQs: workflows for answering multidomain drug discovery questions. Drug Discovery Today **20**(4), 399–405 (2015)
12. Cohen, T., Schvaneveldt, R., Widdows, D.: Reflective Random Indexing and indirect inference: a scalable method for discovery of implicit connections. J. Biomed. Inf. **43**(2), 240–256 (2010)
13. Cohen, T., Whitfield, G.K., Schvaneveldt, R.W., Mukund, K., Rindflesch, T.: EpiphaNet: an interactive tool to support biomedical discoveries. J. Biomed. Discovery Collab. **5**, 21–49 (2010)
14. Devlin, J., Chang, M., Lee, K., Toutanova, K.: BERT: pre-training of deep bidirectional transformers for language understanding. CoRR abs/1810.04805 (2018). http://arxiv.org/abs/1810.04805
15. Eberts, M., Ulges, A.: Span-based joint entity and relation extraction with transformer pre-training (2019)
16. Federhen, S.: The NCBI taxonomy database. Nucleic Acids Res. **40**(D1), D136–D143 (2011)
17. Gopalakrishnan, V., Jha, K., Jin, W., Zhang, A.: A survey on literature based discovery approaches in biomedical domain. J. Biomed. Inform. **93**, 103141 (2019)
18. Hastings, J., et al.: Chebi in 2016: improved services and an expanding collection of metabolites. Nucleic Acids Res. **44**, D1214–D1219 (2015)
19. Hristovski, D., Friedman, C., Rindflesch, T.C., Peterlin, B.: Exploiting semantic relations for literature-based discovery. In: AMIA Annual Symposium proceedings, pp. 349–53 (2006)
20. Hristovski, D., Peterlin, B., Mitchell, J.A., Humphrey, S.M.: Improving literature based discovery support by genetic knowledge integration (2003)
21. Hristovski, D., Rindflesch, T., Peterlin, B.: Using literature-based discovery to identify novel therapeutic approaches. Cardiovasc. hematol. Agents Med. Chem. **11**(1), 14–24 (2013)

22. Hui, W., Lau, W.K.: Application of literature-based discovery in nonmedical disciplines: a survey. In: Proceedings of the 2nd International Conference on Computing and Big Data, ICCBD 2019, pp. 7–11. Association for Computing Machinery, New York (2019)

23. Ijaz, A.Z., Song, M., Lee, D.: MKEM: a multi-level knowledge emergence model for mining undiscovered public knowledge. BMC Bioinform. **11**(Suppl 2), S3 (2010)

24. Kilicoglu, H., Shin, D., Fiszman, M., Rosemblat, G., Rindflesch, T.C.: SemMedDB: a PubMed-scale repository of biomedical semantic predications. Bioinformatics **28**(23), 3158 (2012)

25. Korbel, J.O., et al.: Systematic association of genes to phenotypes by genome and literature mining. PLoS Biol. **3**(5), e134 (2005)

26. Koroleva, A., Kamath, S., Paroubek, P.: Extracting outcomes from articlesreporting randomized controlled trialsusing pre-trained deep language representations. Assisted authoring for avoiding inadequate claims in scientific reporting, chap. 3, pp. 45–68. Print Service Ede, The Netherlands (2019)

27. Koroleva, A., Kamath, S., Paroubek, P.: Measuring semantic similarity of clinical trial outcomes using deep pre-trained language representations. J. Biomed. Inf. X **4**, 100058 (2019)

28. Lee, J., et al.: Biobert: a pre-trained biomedical language representation model for biomedical text mining. arXiv preprint arXiv:1901.08746 (2019)

29. Li, J., et al.: Biocreative V CDR task corpus: a resource for chemical disease relation extraction. Database (2016)

30. Liu, Y., et al.: Roberta: a robustly optimized Bert pretraining approach (2019)

31. Manohar, N., Adam, T., Pakhomov, S., Melton, G., Zhang, R.: Evaluation of herbal and dietary supplement resource term coverage. Stud. Health Technol. Inform. **216**, 785–9 (2015)

32. Mikolov, T., Sutskever, I., Chen, K., Corrado, G., Dean, J.: Distributed representations of words and phrases and their compositionality. In: Proceedings of the 26th International Conference on Neural Information Processing Systems, NIPS 2013, vol. 2, pp. 3111–3119. Curran Associates Inc., Red Hook (2013)

33. Mohan, S., Li, D.: Medmentions: a large biomedical corpus annotated with UMLS concepts. In: Proceedings of the 2019 Conference on Automated Knowledge Base Construction (AKBC 2019) (2019)

34. van Mulligen, E.M., et al.: The EU-ADR corpus: annotated drugs, diseases, targets, and their relationships. J. Biomed. Inform. **45**(5), 879–884 (2012). Text Mining and Natural Language Processing in Pharmacogenomics

35. Neumann, M., King, D., Beltagy, I., Ammar, W.: ScispaCy: fast and robust models for biomedical natural language processing. In: Proceedings of the 18th BioNLP Workshop and Shared Task, pp. 319–327. Association for Computational Linguistics, Florence, Augst 2019

36. Nguyen, N.T., Gabud, R.S., Ananiadou, S.: Copious: a gold standard corpus of named entities towards extracting species occurrence from biodiversity literature. Biodivers. Data J. **7**, e29626 (2019)

37. Ozgür, A., Xiang, Z., Radev, D.R., He, Y.: Literature-based discovery of IFN-gamma and vaccine-mediated gene interaction networks. J. Biomed. Biotechnol. **2010**, 426479 (2010)

38. Papanikolaou, Y., Roberts, I., Pierleoni, A.: Deep bidirectional transformers for relation extraction without supervision. In: Proceedings of the 2nd Workshop on Deep Learning Approaches for Low-Resource NLP, DeepLo 2019 (2019). https://doi.org/10.18653/v1/d19-6108

39. Peters, M., et al.: Deep contextualized word representations. In: Proceedings of the 2018 Conference of the North American Chapter of the Association for Computational Linguistics: Human Language Technologies, vol. 1 (Long Papers) (2018). https://doi.org/10.18653/v1/n18-1202

40. Radford, A., Narasimhan, K., Salimans, T., Sutskever, I.: Improving language understanding with unsupervised learning. Technical report, OpenAI (2018)

41. Rastegar-Mojarad, M., Elayavilli, R.K., Li, D., Prasad, R., Liu, H.: A new method for prioritizing drug repositioning candidates extracted by literature-based discovery. In: IEEE International Conference on Bioinformatics and Biomedicine (BIBM), pp. 669–674. IEEE, November 2015

42. Rindflesch, T.C., Fiszman, M.: The interaction of domain knowledge and linguistic structure in natural language processing: interpreting hypernymic propositions in biomedical text. J. Biomed. Inform. **36**(6), 462–477 (2003). Unified Medical Language System, unified Medical Language System

43. Sang, S., Yang, Z., Wang, L., Liu, X., Lin, H., Wang, J.: SemaTyP: a knowledge graph based literature mining method for drug discovery. BMC Bioinform. **19**(1), 193 (2018)

44. Smith, B., et al.: The obo foundry: coordinated evolution of ontologies to support biomedical data integration. Nat. Biotech. **25**(11), 1251–1255 (2007)

45. Song, M., Han, N.G., Kim, Y.H., Ding, Y., Chambers, T.: Discovering implicit entity relation with the gene-citation-gene network. PloS One **8**(12), e84639 (2013)

46. Swanson, D.R.: Fish oil, Raynaud's syndrome, and undiscovered public knowledge. Perspect. Biol. Med. **30**, 7–18 (1986)

47. Swanson, D.R.: Migraine and magnesium: eleven neglected connections. Perspect. Biol. Med. **31**, 526–557 (1988)

48. Sybrandt, J., Shtutman, M., Safro, I.: MOLIERE: automatic biomedical hypothesis generation system. In: KDD : Proceedings of the International Conference on Knowledge Discovery & Data Mining 2017, pp. 1633–1642, August 2017

49. Torvik, V.I., Smalheiser, N.R.: A quantitative model for linking two disparate sets of articles in MEDLINE. Bioinformatics **23**(13), 1658–1665 (2007)

50. Wilkowski, B., et al.: Graph-based methods for discovery browsing with semantic predications. In: AMIA Annual Symposium Proceedings 2011, pp. 1514–1523 (2011)

51. Williams, A.J., et al.: Open PHACTS: semantic interoperability for drug discovery. Drug Discovery Today **17**(21), 1188–1198 (2012)

52. Wu, H.Y., et al.: An integrated pharmacokinetics ontology and corpus for text mining. BMC Bioinform. **14**, 35 (2013)

53. Zhang, O.R., Zhang, Y., Xu, J., Roberts, K., Zhang, X.Y., Xu, H.: Interweaving domain knowledge and unsupervised learning for psychiatric stressor extraction from clinical notes. In: Benferhat, S., Tabia, K., Ali, M. (eds.) IEA/AIE 2017. LNCS (LNAI), vol. 10351, pp. 396–406. Springer, Cham (2017). https://doi.org/10.1007/978-3-319-60045-1_41

54. Zhang, R., et al.: Exploiting literature-derived knowledge and semantics to identify potential prostate cancer drugs. Cancer Inform **13**(s1), 103–111 (2014). https://doi.org/10.4137/CIN.S13889

Who Is Who in Literature-Based Discovery: Preliminary Analysis

Andrej Kastrin[(✉)] [ID] and Dimitar Hristovski [ID]

Institute for Biostatistics and Medical Informatics, Faculty of Medicine, University of Ljubljana, Ljubljana, Slovenia
{andrej.kastrin,dimitar.hristovski}@mf.uni-lj.si

Abstract. Literature-based discovery (LBD) has undergone an evolution from being an emerging area to a mature research field. Hence it is necessary to summarize the literature and scrutinize general bibliographic characteristics and publication trends. This paper presents very basic scientometric review of LBD in the period 1986–2020. We identified a total of 237 publications on LBD in the Web of Science database. The Journal of Biomedical Informatics published the greatest amount of papers on LBD. The United States plays a leading role in LBD research. Thomas C. Rindflesch is the most productive co-author in the field of LBD. Drawing on these first insights, we aim to better understand the historical progress of LBD in the last 35 years and to be able to improve the publishing practices to contribute to the field in the future.

Keywords: Literature-based discovery · Bibliometric study · Research performance

1 Introduction

Literature-based discovery (LBD) is a text mining approach for automatically generating research hypotheses [11]. LBD is a complex, continually evolving and collaborative research field. To the best of our knowledge, five traditional literature reviews were recently written to elucidate the extent of current knowledge in the LBD research domain. In the same year, Sebastian et al. [10] and Henry et al. [7], published extensive review papers on LBD. The first group of authors provides an in-depth discussion on a broad palette of existing LBD approaches and offers performance evaluations on some recent emerging LBD methodologies. Later authors likewise introduced historical and modern LBD approaches and provided an overview of evaluation methodologies and current trends. Both papers provided the unifying framework for the LBD paradigm, its methodologies, and tools. In 2019 three new review papers appear. Gopalakrishnan et al. [6] provide a more comprehensive analysis of the LBD field; their paper may serves

Supported by the Slovenian Research Agency (Grant No. Z5-9352 and J5-1780).

W. Lu and K. Q. Zhu (Eds.): PAKDD 2020 Workshops, LNAI 12237, pp. 51–59, 2020.
https://doi.org/10.1007/978-3-030-60470-7_6

as a methodological introduction behind particular tools and techniques. Thilakaratne et al. [13] analyzed methodologies used in the LBD using a novel classification scheme and provide a timeline with key milestones in LBD research. In their second paper, Thilakaratne et al. [14] present a large-scale systematic review of the LBD workflow by manually analysing 176 LBD papers. Although these reviews successfully provide insight into the field of LBD through dissecting the research evidence and appropriate classification of research themes, they have not used more sophisticated tools, such as bibliometric and scientometric analysis. Recently, Chen et al. [4] performed first scientometric analysis in the LBD field. They use LBD domain as a proxy to illustrate an intuitive method to compare multiple search strategies in order to identify the most representative body of scientific publications. Consequently, the in-depth analysis in the LBD field is urgently needed, to provide newcomers, researchers, and clinicians with a state-of-the-art scientometric overview of the area.

2 Methods

2.1 Collection of Bibliographic Data

We used Web of Science (WoS) (Clarivate Analytics, Philadelphia, PA, USA) as the data sources for retrieving publications and related metadata in the LBD research domain. In this preliminary analysis, our objective was to include as complete set of publications on LBD as possible without much manual intervention. The search strategy for WoS was defined as: `TS=((`` literature-based discovery'') OR (``undiscovered public knowledge''))`. The time span was from 1986 until 2020. We applied no language, geographic, or any other constraints on the database retrieval procedure. We are aware of at least two limitations of the simple search strategy described above. One limitation is that many conference papers are not indexed in WoS, and therefore, were not included in our analysis. However, we do know there is a considerable number of important LBD papers published in conferences. For example, our group has written at least four well-cited conference papers that are currently not included. The other limitation is that in quite a few cases, the authors have been creative with the titles and abstracts of their LBD papers, and had avoided mentioning the well-established phrases such as *literature-based discovery*. We will address both limitations in our future work by developing a more complex search strategy (or a set of strategies), and by doing various manual interventions.

2.2 Data Analysis

We prepare and summarise statistics on most prolific authors, countries, and journals. We obtain journal metrics including impact factors of the top 10 journals from Journal Citation Reports (Clarivate Analytics, Philadelphia, PA, USA) on January 30, 2020. The main part of the analysis and visualizations were performed in R using the `bibliometrix` package [2].

3   Results

A last search of the databases was performed on January 30, 2020. In further analysis we included a total of 237 bibliographic records. Publications cover a time period of 35 years (1986–2020) beginning with Swanson's first paper on the LBD [12].

The majority of records were original articles ($n = 139$), followed by conference papers ($n = 58$), review papers ($n = 17$), book chapters ($n = 8$), and other material. As of January 30, 2020, the complete set of publications had been cited $n = 5400$ times.

3.1   Publication Evolution over the Years

In the time period 1986–2020, $n = 237$ publications were published about LBD and indexed in WoS. The maximum number of papers ($n = 22$) was published in 2017. It is noteworthy that the term *Literature-Based Discovery* was included in Medical Subject Headings (MeSH) vocabulary in 2013, indicating its high bibliographic importance. Figure 1 depicted the changing pattern of publications (actual and cumulative frequencies) in our data set from 1986 until 2020 for WoS. The reader can observe that the number of publications on LBD increased slowly from 1986 to 2000, but since then it has been increasing significantly. This fact indicates that the field of LBD has acquired significant attention in the last two decades.

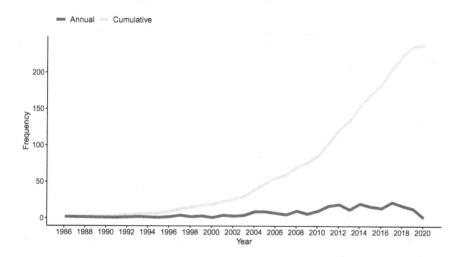

Fig. 1. Number of LBD publications in WoS collection during the period 1986–2020

3.2   Most Productive Authors

Our analysis identifies 530 distinct authors. The majority of the authors write in collaboration with colleagues ($n = 497$). On average we detected 2.24 authors per document and 0.45 documents per author. The authors with the highest number of publications and citations have a tendency to be scientists who drive the research field and have the casting vote for its development. The 10 top authors with the most publications are presented in Table 1. Thomas C. Rindflesch clearly holds the first position with 20 publications, although he is the first author in only one paper on LBD. In addition to raw number of publication, we also present fractionalized number of publications. The author fractionalized number of publications (fNP) is the sum of a unit's publications after assigning each publication the value 1 and dividing the assigned value with the number of authors. Low values in relation to the number of publications indicate a high level of co-authors. For instance, Kostoff achieves high fNP value, because he authored the greatest number of solo publications.

In Fig. 2 we present a co-authorship network of authors in the LBD domain as derived from the WoS database. Although our group has been collaborating in LBD with Thomas C. Rindflesch since the early 2000s, it came to us as a surprise that he is the author with most LBD publications. He is mostly known for the development of SemRep, a natural language processing system that extracts semantic predication from biomedical text [9]. However, Fig. 2 illustrates well that over the years he has collaborated with several other groups, and in the last decade he had his own group publishing in LBD. In network analysis term, he has the highest betweenness centrality and he is the major hub in the co-author network. In the current analysis, we count authorship regardless of the author's position. But most of the time, in most publications, the first author is the one doing most of the work and usually being responsible for the major novelty of the publication. Therefore, as further work, we will create an additional table with the first authors only.

3.3   Most Productive Countries

A total of 34 countries contributed to the selected data set of LBD literature. First, it is worth noting that the LBD production is unevenly distributed across countries. United States commit exactly half of the body of the literature to the treasury of knowledge on LBD ($n = 117$, 50%). This indicates that the US is leading in LBD research. Interestingly, Slovenia, a small country in the heart of Europe, is the second-most productive country with 18 publications (7.6%). Surprisingly, India has no researcher who published about LBD as the first author (Table 2).

3.4   Most Relevant Journals

When analyzing research productivity, it is essential to study the journals in which papers are published. LBD is so specific research field that it has no

Table 1. Top 10 authors based on the total number of publications

Rank	Author	NP	Author	fNP
1	Rindflesch TC	20	Kostoff RN	10.28
2	Kostoff RN	16	Smalheiser NR	7.20
3	Hristovski D	13	Swanson DR	6.37
4	Smalheiser NR	12	Rindflesch TC	4.82
5	Song M	12	Hristovski D	3.62
6	Swanson DR	10	Song M	3.29
7	Kastrin A	9	Kastrin A	2.67
8	Cohen T	7	Preiss J	2.33
9	Persidis A	7	Cohen T	2.20
10	Lee D	6	Ahmed A	2.00

Note: NP = Number of Publications, fNP = fractionalized Number of Publications

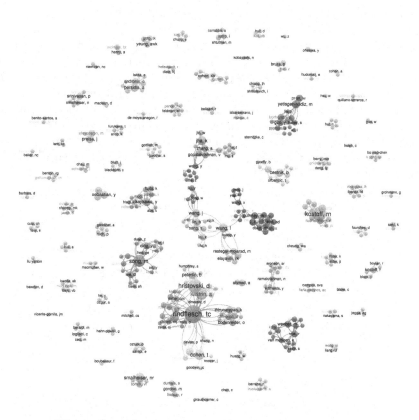

Fig. 2. Co-authorship network of authors on LBD themes

Table 2. Top 10 productive countries for LBD research

Rank	Country	NP	Prop	SCP	MCP	MCPr
1	USA	117	0.50	98	19	0.162
2	Slovenia	18	0.08	7	11	0.611
3	China	17	0.07	15	2	0.118
4	Korea	13	0.06	8	5	0.385
5	United Kingdom	12	0.05	8	4	0.333
6	Japan	11	0.05	9	2	0.182
7	Australia	5	0.02	3	2	0.400
8	Netherlands	5	0.02	3	2	0.400
9	Spain	5	0.02	5	0	0.000
10	Canada	4	0.02	3	1	0.250

Note: NP = Number of Publications, Prop = Proportion of Publications, SCP = Single Country Publications, MCP = Multiple Country Publications

specialized journals. Instead, the LBD research is published mainly in journals related to (biomedical) informatics and bioinformatics. Table 3 summarizes the details about the top 10 journals. Not surprisingly, with respect to the number of publications, *Journal of Biomedical Informatics* had published 14 papers on LBD research, followed by *Technological Forecasting and Social Change* with 11 published articles. *Briefings in Bioinformatics*, which has the highest impact factor in our list, has published only 6 papers on LBD. Out of the 10 journals, the majority are published in the United States. Journals from the top 10 list publish LBD papers from the beginning of the 2000s, with the exception of the *Journal of the American Society for Information Science* which was active between the years 1987 and 1999.

Table 3. Journals with most LBD publications

Rank	Source	NP	IF
1	J. Biomed. Inform.	14	2.950
2	Technol. Forecast. Soc. Chang.	11	3.815
3	J. Am. Soc. Inf. Sci. Technol.	10	2.452
4	BMC Bioinformatics	7	2.511
5	Bioinformatics	6	4.531
6	Brief. Bioinform.	6	9.101
7	J. Am. Med. Inf. Assoc.	6	4.292
8	PLoS One	6	2.776
9	Scientometrics	6	2.770
10	J. Doc.	5	1.573

Note: NP = Number of Publications, IF = Impact Factor

3.5   Publication Hallmarks

By employing the processed bibliometric data, we can identify the most important hallmarks of LBD research. The top 10 most cited papers are listed in Table 4, including their first author, year of publication, journal, the total number of citations and number of citations per year. Data are ranked by the number of citations. Swanson is the author of five listed publications including his seminal paper on fish oil and Raynaud's disease which is the first on the list [12]. The second most cited paper is a review article published by Cohen et al. [5] in which they discuss various text mining approaches including automated hypothesis generation from literature. Swanson's paper is categorically the first hallmark of LBD research. However, it is important to note that Cohen's paper has more than two-times more citations per year. This is probably due to the high impact factor of the journal in which the paper was published and because of its interestingness for the broader domain of researchers. These ten publications covered the theoretical research as well as practical applications of LBD. However, all these papers were published before 2005, although important scientific achievements in LBD were published also later on.

Table 4. The top 10 papers in the LBD domain based on the number of citations

Rank	Paper	TC	TCY
1	Swanson DR, 1986, Perspect Biol Med	402	11.82
2	Cohen AM, 2005, Brief Bioinform	363	24.20
3	Dumais ST, 2004, Annu Rev Inform Sci	290	18.12
4	Swanson DR, 1997, Artif Intell	230	10.00
5	Swanson DR, 1986, Libr Quart	169	4.97
6	Srinivasan P, 2004, J Am Soc Inf Sci Tec	155	9.69
7	Kostoff RN, 2004, Technol Forecast Soc	150	9.38
8	Hristovski D, 2005, Int J Med Inform	140	9.33
9	Zweigenbaum P, 2007, Brief Bioinform	131	10.08
10	Weeber M, 2001, J Am Soc Inf Sci Tec	126	6.63

Note: TC = Total Citations, TCY = Total Citations per Year

4   Discussion

Through very basic scientometric analysis, this study aimed to reveal worldwide scientific productivity and research trends in LBD over the last three decades (1986–2020). To the best of our knowledge, this paper, although in its preliminary version, is the first scientometric analysis in the field of LBD.

Understanding the past and current body of publications is the sine qua non for the advancement of LBD research in the future. In the last decade, a plethora of studies has been published examining knowledge structure and evolution through the bibliographic lens of particular scientific fields. The lack of a similar study in the LBD area makes it difficult if not impossible to compare LBD with other research fields. However, LBD is inherently lean to biomedicine and to medical informatics in particular. There are two reasons for this fact. First, historically, LBD originates from the medical applications. Second, practically, MEDLINE distribution is freely available to researchers that is not the case with Scopus or WoS.

A conspicuous change in the number of papers published per year suggests that a major turning point is occurring in the field. We found that the number of publications increased over the last 20 years, particularly since 2000. The development of the LBD field is associated with great progress in computer science and natural language processing in particular. The total number of citations accumulate over the years and consequently, the recent papers do not have enough time to acquire more citations. However, the growth of publications and citations in recent years indicates a promising future of LBD.

Scientific productivity is strongly correlated with international collaboration among researchers, countries, and institutions [8]. Studies investigating the scientific impact of cross-institution groups confirmed that their papers have a higher citation rate in comparison to papers produced by a single research group. Papers with international co-authorship have an even higher impact [15]. Most of the research produced in the field of LBD is generated in the cliques of researchers. Even though the collaboration and internationalization among researchers have certain downsides, it provides great benefits. Abramo et al. [1] demonstrated an increasing trend in collaboration among institutions that could be attributed to different policies stimulating research collaboration (e.g., the EU Framework Programme for Research and Innovation). We are aware of at least one successful EU FP7 funded project from the domain of LBD named BISON (2008–2011) that investigates novel methods for discovering new, domain bridging connections and patterns from heterogeneous data sources [3].

For further work, we intend to greatly expand the analysis to the Scopus, Pubmed, Google Scholar, and Dimensions databases. To build an universum of relevant publications, we will employ a strategy that combines regular query search with cascading citation expansion approach as proposed recently by Chen et al. [4]. Preliminary work reveals that such expansion improves the results a lot. Next, we already work on the science mapping of the LBD domain. When completed, this study will elucidate a past, present, and future image of LBD in great detail.

References

1. Abramo, G., D'Angelo, C.A., Solazzi, M.: The relationship between scientists' research performance and the degree of internationalization of their research. Scientometrics **86**(3), 629–643 (2011)
2. Aria, M., Cuccurullo, C.: Bibliometrix: an R-tool for comprehensive science mapping analysis. J. Inf. **11**(4), 959–975 (2017)

3. Berthold, M.R. (ed.): Bisociative Knowledge Discovery. LNCS (LNAI), vol. 7250. Springer, Heidelberg (2012). https://doi.org/10.1007/978-3-642-31830-6
4. Chen, C., Song, M.: Visualizing a field of research: a methodology of systematic scientometric reviews. PloS One **14**(10), e0223994 (2019)
5. Cohen, A.M., Hersh, W.R.: A survey of current work in biomedical text mining. Brief. Bioinform. **6**(1), 57–71 (2005)
6. Gopalakrishnan, V., Jha, K., Jin, W., Zhang, A.: A survey on literature based discovery approaches in biomedical domain. J. Biomed. Inform. **93**, 103141 (2019)
7. Henry, S., McInnes, B.: Literature based discovery: models, methods, and trends. J. Biomed. Inform. **74**, 20–32 (2017)
8. Lee, S., Bozeman, B.: The impact of research collaboration on scientific productivity. Soc. Stud. Sci. **35**(5), 673–702 (2005)
9. Rindflesch, T.C., Fiszman, M.: The interaction of domain knowledge and linguistic structure in natural language processing: interpreting hypernymic propositions in biomedical text. J. Biomed. Inform. **36**(6), 462–477 (2003)
10. Sebastian, Y., Siew, E.G., Orimaye, S.: Emerging approaches in literature-based discovery: techniques and performance review. Knowled. Eng. Rev. **32**, E12 (2017)
11. Smalheiser, N.: Rediscovering Don Swanson: the past, present and future of literature-based discovery. J. Data Inf. Sci. **2**(4), 43–64 (2017)
12. Swanson, D.: Fish oil, Raynaud's syndrome, and undiscovered public knowledge. Perspect. Biol. Med. **30**(1), 7–18 (1986)
13. Thilakaratne, M., Falkner, K., Atapattu, T.: A systematic review on literature-based discovery: general overview, methodology, & statistical analysis. ACM Comput. Surv. (CSUR) **52**(6), 1–34 (2019)
14. Thilakaratne, M., Falkner, K., Atapattu, T.: A systematic review on literature-based discovery workflow. PeerJ Comput. Sci. **5**, e235 (2019)
15. Thonon, F., et al.: Measuring the outcome of biomedical research: a systematic literature review. PLoS One **10**(4), e0122239 (2015)

Workshop on Data Science for Fake News (DSFN 2020)

Adversarial Deep Factorization
for Recommender Systems

Ziqing Chen[1](✉), Yiwei Zhang[2], and Zongyang Li[3]

[1] University of Waterloo, Waterloo, Canada
z443chen@uwaterloo.ca
[2] Kings College London, London, UK
yiwei.1.zhang@kcl.ac.uk
[3] The University of Sydney, Sydney, Australia
zoli7177@uni.sydney.edu.au

Abstract. Recently, hacking from malicious users is a big challenge for major corporations and government organizations. Tensor factorization methods have been developed to learn predictive models in multi-criteria recommender systems by dealing with the three-dimensional (3D) user-item-criterion ratings. However, they suffer from the data sparsity and contamination issues in real applications. In order to overcome these problems, we propose a general architecture of adversarial deep factorization (ADF) by integrating deep representation learning and tensor factorization, where the side information is embedded to provide an effective compensation for tensor sparsity, and the adversarial learning is adopted to enhance the model robustness. Experimental results on three real-world datasets demonstrate that our ADF schemes outperform state-of-the-art methods on multi-criteria rating predictions. Specifically, the proposed model considers a brilliant combination with adversarial stacked denoising autoencoder (ASDAE), where the adversarial training is used to learn effective latent factors instead of being placed on the extrinsic rating inputs.

1 Introduction

Recently, hacking from malicious users is a big challenge for major corporations and government organizations [3]. The fakeness serious adverse effects to scare or harm the society [2]. Recommendation system has become one of the focuses of malicious users' intentional attacks. Recommender systems are capable of assisting users in identifying items that fit their own tastes best, and helping commerce companies to target customers. With the data explosion in recent years, recommender systems have become increasingly attractive. Traditional single-criterion recommender systems typically operate on the two-dimensional (2D) user-item ratings [6,9,14]. Nevertheless, they cannot work well for multi-criteria recommender systems that contains multiple criterion-specific ratings.

With the emergence of multimodal or multiaspect data, multi-criteria recommender systems become more and more important. The list of applications ranges from social network analysis to brain data analysis, and from web

© Springer Nature Switzerland AG 2020
W. Lu and K. Q. Zhu (Eds.): PAKDD 2020 Workshops, LNAI 12237, pp. 63–71, 2020.
https://doi.org/10.1007/978-3-030-60470-7_7

mining and information retrieval to healthcare analytic [16]. Especially, multi-criteria recommendation is requisite in a variety of online e-commerce websites and traveling portals. There have been many significant efforts made to deal with multi-criteria recommendations [4]. Existing techniques can be classified into three categories: heuristic neighborhood-based approaches [11], aggregation-based approaches [12], and model-based approaches [17].

Fig. 1. Example from TripAdvisor

Nevertheless, all prior techniques for multi-criteria recommendations suffer from the data sparsity and contamination problems. In other words, as the rating tensor is very sparse and contains the malicious users' fake information in real applications, the resulting latent factors are still not good enough to achieve a satisfactory performance.

Recommender systems are vulnerable to malicious users who make some fake information to deviate the original output, causing a mistake in recommendations [15]. Recently, adversarial machine learning techniques have shown a great performance on various tasks such as image captioning, sequence generation, image-to-image translation, neural machine translation, and information retrieval. Despite a great potential being shown, there is little work about the application of adversarial machine learning techniques in recommender systems (e.g. [8,19]). Wang et al. [19] propose a minimax adversarial game to reduce the susceptibility of traditional trained models to adversarial examples, which becomes the state-of-the-art method in three different tasks, including web search, question answering and item recommendation. He et al. [8] improve information retrieval GAN (IRGAN) by choosing a model with a pairwise loss function and then inducing the adversarial noise into the training process. Different from prior works, we further leverage the benefits of incorporating the side information into decomposition to solve the data sparsity in multi-criteria recommendations, where the adversarial training is considered to solve the data contamination.

In order to overcome the data sparsity and contamination problems in multi-criteria recommender systems, we attempt to incorporate auxiliary information into the rating tensor to exploit prior features, and using adversarial learning to defense the attack from fake information. As shown in Fig. 1, the information of customers belongs to the user's auxiliary information while the information of hotels belongs to the item's auxiliary information.

In this paper, we propose a general architecture of adversarial deep factorization (ADF) by integrating deep representation learning and tensor factorization, where the side information is embedded to provide an effective compensation for tensor sparsity, and the adversarial learning is used to enhance the model robustness. We exhibit a specific instantiation of our architecture by combining adversarial stacked denoising autoencoder (ASDAE) and CANDECOMP/PARAFAC (CP) tensor factorization, where the side information of both users and items is tightly coupled with the sparse multi-criteria ratings and the latent factors are learned based on the joint optimization.

However, the challenge lies in how to effectively incorporate the side information into the rating tensor and how to take advantage of adversarial learning to enhance the robustness of the model. To overcome these two challenges, we propose an adversarial deep factorization (ADF) method in this paper, which integrates deep representation learning and tensor factorization. *On the one hand*, the side information of users and items is encoded by stacked denoising autoencoder (SDAE) to respectively compensate the users and items' latent factors from tucker decomposition. Three latent factor matrices are updated using the gradient method. *On the other hand*, the adversarial training is used to learn effective latent factors at middle layer instead of being placed on the extrinsic input of SDAE.

The contributions of this paper can be summarized as follows

- To solve the data sparsity and contamination problems in multi-criteria recommendations, we propose a generic architecture to integrate deep structure and tensor factorization, which is the first work to abundantly incorporate the side information and defense malicious users' fake information in multi-criteria recommender systems;
- We present a specific ADF scheme where CP tensor factorization is combined with two ASDAEs for both users and items to improve tensor factorization based recommendations;
- Experiment results on three real datasets demonstrate that our proposed ADF outperforms state-of-the-art methods in terms of multi-criteria recommendations.

2   Adversarial Deep Factorization

The specific ADF scheme is composed of three components: a ASDAE for users, a ASDAE for items, and tensor factorization, as shown in Fig. 2. In the proposed

ADF, an individual ASDAE takes both the real and fake side information as input.

Considering the ASDAE for users in Fig. 2, the representation $h_l^{(u)}$ at each hidden layer can be obtained as

$$\mathbf{h}_l^{(u)} = g\left(\mathbf{W}_l^{(u)}\mathbf{h}_{l-1}^{(u)} + \mathbf{b}_l^{(u)}\right) \tag{1}$$

and the output at the layer $L^{(u)}$ can be obtained as

$$\hat{\mathbf{m}} = f\left(\mathbf{W}_{L^{(u)}}^{(u)}\mathbf{h}_{L^{(u)}}^{(u)} + \mathbf{b}_{L^{(u)}}^{(u)}\right), \tag{2}$$

where $l \in \{1, 2, \cdots, L^{(u)} - 1\}$; $g(\cdot)$ and $f(\cdot)$ are activition functions for the hidden and output layers. The side information $\hat{\mathbf{m}}$ is the input to the first layer, $\mathbf{h}_r^{(u)}$ denotes deep representations from the middle layer and $\hat{\mathbf{m}}$ denotes the output of the users' ASDAE. Similar results can be obtained for the items' ASDAE by replacing the superscript (u) with (v) and replacing $\mathbf{m}$ with $\mathbf{n}$.

As observed in Fig. 2, the ASDAEs take the side information of users and items as input to learn the latent representations $\mathbf{h}_r^{(u)}$ and $\mathbf{h}_r^{(v)}$ that are used to compensate the latent factor vectors $\mathbf{u}$ and $\mathbf{v}$ in tensor factorization.

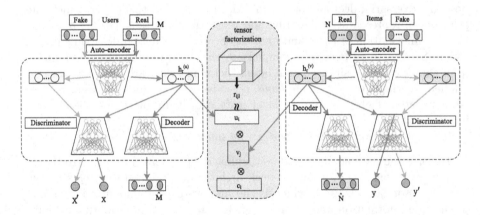

Fig. 2. The structure of the proposed ADF

Loss Function. The proposed ADF learns users' latent factors, items' latent factors and criteria latent factors from both the rating tensor and the side information through the following objective function

$$\min_{\Theta} \mathcal{J} = \frac{1}{2}\left(\mathcal{L}_t + \mathcal{L}_r + \mathcal{L}_a + \lambda f_{reg} + \mathcal{L}_d\right), \tag{3}$$

where the overall loss function $\mathcal{J}$ consists of five components: the loss of tensor factorization $\mathcal{L}_t$, the reconstruction cost of the side information $\mathcal{L}_r$, the approximation error between deep representation and latent factors $\mathcal{L}_a$, the construction cost of adversarial training $\mathcal{L}_d$, and the regularization term f_{reg} that prevents overfitting.

The first term $\mathcal{L}_t$ denotes the loss of factorization on a sparse rating tensor

$$\min_{\theta_t} \mathcal{L}_t = \|\mathbf{I} \odot (\mathbf{R} - \mathbf{U} \otimes \mathbf{V} \otimes \mathbf{C})\|^2, \tag{4}$$

where $\theta_t = \{\mathbf{U}, \mathbf{V}, \mathbf{C}\}$; the binary tensor $\mathbf{I}$ is an indicator of sparsity, in which each element indicates whether the corresponding rating is observed $(= 1)$ or not $(= 0)$; $\otimes$ means the outer product of latent factor *vectors* in the corresponding matrix as mentioned above; and $\odot$ is the element-wise operation.

Secondly, the reconstruction cost of the side information for both users and items can be expressed as

$$\min_{\theta_r} \mathcal{L}_r = \alpha \sum_i (\mathbf{m}_i - \hat{\mathbf{m}}_i)^2 + \beta \sum_j (\mathbf{n}_j - \hat{\mathbf{n}}_j)^2, \tag{5}$$

where $\theta_r = \{\mathbf{W}^u, \mathbf{b}^u, \mathbf{W}^v, \mathbf{b}^v\}$, α and β are penalty parameters.

Furthermore, the approximation error between deep representation and latent factor vectors for both users and items can be expressed as

$$\min_{\theta_a} \mathcal{L}_a = \rho \sum_i \left(\mathbf{u}_i - \mathbf{h}_r^{(u_i)}\right)^2 + \gamma \sum_j \left(\mathbf{v}_j - \mathbf{h}_r^{(v_j)}\right)^2, \tag{6}$$

where $\theta_a = \{\mathbf{W}^u, \mathbf{b}^u, \mathbf{W}^v, \mathbf{b}^v, \mathbf{U}, \mathbf{V}\}$, ρ and γ are penalty parameters.

And the loss function of adversarial training can be expressed as

$$\min_{\theta_r} \mathcal{L}_d = \left\|\sum_i log(x_i) + log(1 - x_i')\right\|^2$$
$$+ \left\|\sum_j log(y_j) + log(1 - y_j')\right\|^2, \tag{7}$$

where $\theta_r = \{\mathbf{W}^u, \mathbf{b}^u, \mathbf{W}^v, \mathbf{b}^v\}$, x_i and x_i' is the probability of real and fake side information of users, and y_i and y_i' is the probability of real and fake side information of items.

And the last term denotes the regularization term f_{reg} which is formulated as follows

$$f_{reg} = \sum_i \|\mathbf{u}_i\|^2 + \sum_j \|\mathbf{v}_j\|^2 + \|\mathbf{W}^{(u)}\|^2 + \|\mathbf{W}^{(v)}\|^2 + \|\mathbf{b}^{(u)}\|^2 + \|\mathbf{b}^{(v)}\|^2 \tag{8}$$

and the overall $\Theta = \theta_t \cup \theta_r \cup \theta_a$ in (3).

Optimization. To solve this problem, the alternative optimization algorithm is considered by utilizing the following two-step procedure.

Step I: Given all weights $\mathbf{W}$ and biases $\mathbf{b}$, the gradients of the overall loss function $\mathcal{J}$ in (3) with respect to $\mathbf{u}_i$, $\mathbf{v}_j$, $\mathbf{c}_l$, can be obtained as

$$\frac{\partial \mathcal{J}}{\partial \mathbf{u}_i} = -\sum_j \sum_l I_{ijl} \left(r_{ijl} - \mathbf{u}_i \mathbf{v}_j \mathbf{c}_l \right) \left(\mathbf{v}_j \mathbf{c}_l \right) + \rho \left(\mathbf{u}_i - \mathbf{h}_r^{(u_i)} \right) + \lambda \mathbf{u}_i,$$

$$\frac{\partial \mathcal{J}}{\partial \mathbf{v}_j} = -\sum_i \sum_l I_{ijl} \left(r_{ijl} - \mathbf{u}_i \mathbf{v}_j \mathbf{c}_l \right) \left(\mathbf{u}_i \mathbf{c}_l \right) + \gamma \left(\mathbf{v}_j - \mathbf{h}_r^{(v_j)} \right) + \lambda \mathbf{v}_j,$$

$$\frac{\partial \mathcal{J}}{\partial \mathbf{c}_l} = -\sum_i \sum_j I_{ijl} \left(r_{ijl} - \mathbf{u}_i \mathbf{v}_j \mathbf{c}_l \right) \left(\mathbf{u}_i \mathbf{v}_j \right) + \lambda \mathbf{c}_l, \tag{9}$$

where I_{ijl} indicates whether the corresponding rating is observed ($=1$) or not ($=0$).

Step II: Fixed the latent factors $\mathbf{U}$, $\mathbf{V}$ and $\mathbf{C}$, all weights $\mathbf{W}$ and biases $\mathbf{b}$ of both ASDAEs can be learned by backpropagation with SGD method

$$\frac{\partial \mathcal{J}}{\partial \mathbf{W}^{(u)}} = -\rho \sum_i \left(\mathbf{u}_i - \mathbf{h}_r^{(u_i)} \right) \frac{\partial \mathbf{h}_r^{(u_i)}}{\partial \mathbf{W}^{(u)}} + \alpha \sum_i \left(\mathbf{m}_i - \hat{\mathbf{m}}_i \right) \frac{\partial \hat{\mathbf{m}}_i}{\partial \mathbf{W}^{(u)}}$$
$$- \left(\frac{1}{\mathbf{x}} \right) \frac{\partial \mathbf{x}}{\partial \mathbf{W}^{(u)}} + \left(\frac{1}{1 - \mathbf{x}'} \right) \frac{\partial \mathbf{x}'}{\partial \mathbf{W}^{(u)}} + \lambda \mathbf{W}^{(u)},$$

$$\frac{\partial \mathcal{J}}{\partial \mathbf{W}^{(v)}} = -\gamma \sum_j \left(\mathbf{v}_j - \mathbf{h}_r^{(v_j)} \right) \frac{\partial \mathbf{h}_r^{(v_j)}}{\partial \mathbf{W}^{(v)}} + \beta \sum_j \left(\mathbf{n}_j - \hat{\mathbf{n}}_j \right) \frac{\partial \hat{\mathbf{n}}_j}{\partial \mathbf{W}^{(v)}}$$
$$- \left(\frac{1}{\mathbf{y}} \right) \frac{\partial \mathbf{y}}{\partial \mathbf{W}^{(u)}} + \left(\frac{1}{1 - \mathbf{y}'} \right) \frac{\partial \mathbf{y}'}{\partial \mathbf{W}^{(u)}} + \lambda \mathbf{W}^{(v)}, \tag{10}$$

and $\frac{\partial \mathcal{J}}{\partial \mathbf{b}^{(u)}}$ and $\frac{\partial \mathcal{J}}{\partial \mathbf{b}^{(v)}}$ can be obtained by replacing $\mathbf{W}$ with $\mathbf{b}$ in (10). Iterate two steps until the convergence.

3   Experiments

3.1   Datasets and Evaluation Metric

To evaluate the algorithms, we use three public datasets, two from TripAdvisor and one from Yahoo!Movie. All the datasets are commonly used for evaluating the performance of recommender systems [10]. In experiments, five-fold cross validation was applied to each dataset. We use the root mean squared error (RMSE) and the mean absolute error (MAE) [7] as the evaluation metric.

Table 1. Performance comparison in terms of RMSE.

Algorithm	TripAdvisor-12M			TripAdvisor-20M			Yahoo!Movie		
	60%	80%	95%	60%	80%	95%	60%	80%	95%
AFBM [1]	1.178	1.053	1.045	1.219	1.167	1.096	1.173	1.165	1.152
CMF [18]	1.274	1.058	1.038	1.184	1.140	1.130	1.038	1.069	0.888
HCF [5]	1.089	1.066	1.016	1.128	1.073	1.030	0.933	0.834	0.827
DCF [13]	1.163	1.069	1.031	1.164	1.094	1.036	0.968	0.917	0.874
t-SVD [20]	1.151	1.040	0.961	1.181	1.075	1.039	0.873	0.829	0.804
ADF	**1.022**	**1.016**	**0.869**	**1.082**	**1.049**	**1.029**	**0.750**	**0.647**	**0.596**

Table 2. Performance comparison in terms of MAE.

Algorithm	TripAdvisor-12M			TripAdvisor-20M			Yahoo!Movie		
	60%	80%	95%	60%	80%	95%	60%	80%	95%
AFBM	0.834	0.828	0.822	0.814	0.807	0.757	0.908	0.903	0.896
CMF	0.889	0.786	0.751	0.803	0.793	0.782	0.774	0.787	0.672
HCF	0.748	0.746	0.738	0.741	0.720	0.710	0.703	0.614	0.613
DCF	0.827	0.758	0.751	0.870	0.775	0.736	0.726	0.692	0.653
t-SVD	0.808	0.784	0.750	0.751	0.712	0.708	0.687	0.660	0.640
ADF	**0.681**	**0.675**	**0.607**	**0.725**	**0.706**	**0.703**	**0.545**	**0.467**	**0.427**

3.2 Summary of Experimental Results

We evaluate our proposed ADF on three datasets in comparison to state-of-the-art recommendation baselines.

Table 1 illustrates the comparison results in terms of the average RMSE while Table 2 illustrates the comparison results in terms of the average MAE, where the lowest RMSE/MAE of each dataset is highlighted in boldface. The proposed ADF clearly outperforms the well-established baselines in terms of RMSE/MAE, achieving the *best* performance for all cases.

Furthermore, it is observed that HCF, DCF and CMF outperform AFBM in general cases, which demonstrates the effectiveness of incorporating the side information in either 2D rating matrix or 3D rating tensor. That ADF, HCF and DCF outperform CMF indicates that deep structure can acquire better features of the side information. HCF, DCF, CMF and AFBM only consider the correlation between arbitrary two of three dimensions so ADF and t-SVD outperform these methods. That ADF outperform DCF and t-SVD indicates the effectiveness of adversarial training.

4   Conclusion

In this paper, to defense the attack from fake information, we have proposed a general architecture and a specific instantiation of adversarial deep factorization (ADF) for recommendation. Our experimental results on real-world data sets have demonstrated that ADF outperforms state-of-art models. As part of future work, we will consider to improve the way of the incorporation of the side information to pursue a better performance in multi-criteria recommender systems.

References

1. Adomavicius, G., Kwon, Y.: New recommendation techniques for multicriteria rating systems. IEEE Intell. Syst. **22**(3), 48–55 (2007)
2. Anoop, K., Deepak, P., Lajish, V.: Emotion cognizance improves fake news identification. arXiv preprint arXiv:1906.10365 (2019)
3. Chakraborty, T., Jajodia, S., Katz, J., Picariello, A., Sperli, G., Subrahmanian, V.: FORGE: a fake online repository generation engine for cyber deception. IEEE Trans. Dependable Secure Comput. (2019)
4. Chen, Z., Gai, S., Wang, D.: Deep tensor factorization for multi-criteria recommender systems. In: IEEE International Conference on Big Data, Big Data 2019, Los Angeles, CA, USA, 9–12 December 2019 (2019)
5. Dong, X., Yu, L., Wu, Z., Sun, Y., Yuan, L., Zhang, F.: A hybrid collaborative filtering model with deep structure for recommender systems. In: AAAI, pp. 1309–1315 (2017)
6. Gai, S., Zhao, F., Kang, Y., Chen, Z., Wang, D., Tang, A.: Deep transfer collaborative filtering for recommender systems. In: Nayak, A.C., Sharma, A. (eds.) PRICAI 2019. LNCS (LNAI), vol. 11672, pp. 515–528. Springer, Cham (2019). https://doi.org/10.1007/978-3-030-29894-4_42
7. He, X., Chen, T., Kan, M.Y., Chen, X.: TriRank: review-aware explainable recommendation by modeling aspects. In: Proceedings of the 24th ACM International on Conference on Information and Knowledge Management, pp. 1661–1670. ACM (2015)
8. He, X., He, Z., Du, X., Chua, T.S.: Adversarial personalized ranking for recommendation. In: The 41st ACM SIGIR, pp. 355–364. ACM (2018)
9. Jahrer, M., Töscher, A., Legenstein, R.: Combining predictions for accurate recommender systems. In: Proceedings of the 16th ACM SIGKDD International Conference on Knowledge Discovery and data Mining, pp. 693–702. ACM (2010)
10. Jannach, D., Zanker, M., Fuchs, M.: Leveraging multi-criteria customer feedback for satisfaction analysis and improved recommendations. Inf. Technol. Tourism **14**(2), 119–149 (2014). https://doi.org/10.1007/s40558-014-0010-z
11. Lakiotaki, K., Matsatsinis, N.F., Tsoukias, A.: Multicriteria user modeling in recommender systems. IEEE Intell. Syst. **26**(2), 64–76 (2011)
12. Lakiotaki, K., Tsafarakis, S., Matsatsinis, N.: UTA-Rec: a recommender system based on multiple criteria analysis. In: Proceedings of the 2008 ACM Conference on Recommender Systems, RecSys 2008, pp. 219–226. ACM, New York (2008). http://doi.acm.org/10.1145/1454008.1454043

13. Li, S., Kawale, J., Fu, Y.: Deep collaborative filtering via marginalized denoising auto-encoder. In: Proceedings of the 24th ACM International on Conference on Information and Knowledge Management, CIKM 2015, pp. 811–820. ACM, New York (2015). http://doi.acm.org/10.1145/2806416.2806527
14. Mnih, A., Salakhutdinov, R.R.: Probabilistic matrix factorization. In: Advances in Neural Information Processing Systems, pp. 1257–1264 (2008)
15. O'Mahony, M., Hurley, N., Kushmerick, N., Silvestre, G.: Collaborative recommendation: a robustness analysis. ACM TOIT 4(4), 344–377 (2004)
16. Papalexakis, E.E., Faloutsos, C., Sidiropoulos, N.D.: Tensors for data mining and data fusion: models, applications, and scalable algorithms. ACM Trans. Intell. Syst. Technol. 8(2), 1–44 (2016). http://dl.acm.org/citation.cfm?doid=3004291.2915921
17. Sahoo, N., Krishnan, R., Duncan, G., Callan, J.: Research note-the halo effect in multicomponent ratings and its implications for recommender systems: the case of Yahoo! Movies. Inf. Syst. Res. 23(1), 231–246 (2011). https://pubsonline.informs.org/doi/abs/10.1287/isre.1100.0336
18. Singh, A.P., Gordon, G.J.: Relational learning via collective matrix factorization. In: KDD, pp. 650–658 (2008)
19. Wang, J., et al.: IRGAN: a minimax game for unifying generative and discriminative information retrieval models. In: Proceedings of the 40th ACM SIGIR, pp. 515–524. ACM (2017)
20. Zhang, Z., Aeron, S.: Exact tensor completion using t-SVD. IEEE Trans. Signal Processing 65(6), 1511–1526 (2017)

Detection of Spammers Using Modified Diffusion Convolution Neural Network

Hui Li, Wenxin Liang$^{(\boxtimes)}$, and Zihan Liao

School of Software, Dalian University of Technology, Dalian 116620, China
lh_wyi@yeah.net, wxliang@dlut.edu.cn, elvis@mail.dlut.edu.cn

Abstract. Social network brings convenience to our life, but it also provides a platform for spammers to spread malicious information and links. Most of the existing methods for identifying spammers mainly rely on the user's behavior information to learn classification models. But, privacy and information security issues make it impossible to monitor all behavioral of users. In addition, owing to the diversity and variability of spammers' strategies, it is difficult to distinguish them from legitimate users only by their own behavior. To solve this challenge, in this paper we propose a novel spammer detecting method using DCNN (Diffusion Convolution Neural Network) which is a graph-based model. And DCNN model can learn behavior information from other users through the graph structure (i.e., social network relationships). However, the original DCNN model is a general classification model. In order to make the original DCNN model more effective for detecting spammers, we modify it by using attenuation Coefficient and social Regularization, which is called DCNN+CR model. The experimental results on real-world Twitter dataset show that the proposed DCNN+CR model outperforms existing methods, especially in terms of accuracy and F1-score.

Keywords: Social network · Spammer · Diffusion · Graph structure

1 Introduction

Online social networks, such as Twitter and Facebook, have become popular social platforms where users can interact with friends, post real-time messages and share the fun in their lives [6,10,11]. Unfortunately, spammers use these platforms to phish scams and spread illegal advertisements and pornography, etc. [14,15]. These malicious behaviors of spammers have not only affected legitimate user, but also seriously hindered the development of social networks [12]. Therefore, it is very meaningful to design effective spammer detection methods.

In order to solve the above problem, researchers have done a lot of researches on spammer like using the significant features of users' behavior and suitable

This work was partially supported by the National Science Foundation of China (No. 61632019) and the Fundamental Research Funds for the Central Universities (No. DUT19RC(3)048).

W. Lu and K. Q. Zhu (Eds.): PAKDD 2020 Workshops, LNAI 12237, pp. 72–79, 2020.
https://doi.org/10.1007/978-3-030-60470-7_8

detection algorithms to distinguish spammers [2,8,13]. However, with the development of spammer strategies, those methods which only depend on user's own features could not effectively detect social spammers. In addition, privacy and information security issues are becoming more and more important, which makes it more difficult to obtain all behavior information of users on the website. Therefore, some researchers have proposed to use social network relationships(e.g., social regularization) to optimize the detection methods of spammer [5,6,16]. It achieved better results compared to previous methods. However, these optimization methods do not enrich and optimization the behavior features of users.

In this paper, we propose to use the DCNN+CR for detecting spammers. It can learn the behavior information diffused from other users through social network relationships. Combined with behavior information learned from other users, we can extract more useful information for spammer detection. The original DCNN is a graph-based general model. Therefore, for specific problems such as detection of spammers, we appropriately modify the DCNN model by attenuation coefficient and social regularization, named DCNN+CR. In summary, the contributions of this paper are as follows:

- Through the DCNN model, we propose a new way to use social network relationships.
- Problem detection for spammers, we optimize the general DCNN model through attenuation coefficient and social regularization, named DCNN+CR.
- Through the experimental results, we get that the combined features (i.e., combining multiple kinds of features into the same feature matrix) play different effects in different models.

The remainder of this paper is organized as follows: Sect. 2 definite the problems and introduce our DCNN+CR model. Section 3 demonstrates our evaluation process and experimental results. We conclude the discussion in Sect. 4.

2 Problem Definition and Model Introduction

In this section, we introduce the terminology involved in this paper and then formally define the problem of detecting spammers. Finally, we will introduce the basic DCNN and our DCNN+CR model.

2.1 Problem Definition

Definition 1. *The features information X is a matrix of $N \times F$, where N is the number of users, and F is the number of features.*

Definition 2. *Social network information $R = \{V, E\}$ that is composed of vertices V and edges E denotes users' following relationships. For example, user v_i follow user v_j, i.e., there is a direct edge from v_i to v_j.*

Definition 3. *Y is the identify labeled matrix of $N \times c$, where c represents the number of classes. Here are mainly divided into legitimate users and spammers. If user u_i is a spammer, $Y_i = (0, 1)$; otherwise, $Y_i = (1, 0)$.*

Problem 1. Given the feature information X, social network information R, and labeled information Y, our task is to learn a model to automatically identify unknown users as spammers or legitimate users.

2.2 DCNN Model

DCNN model is a graph convolutional neural combined with graph structure data. Comparing with the standard convolution neural network, DCNN has no pooling operation. The DCNN model mainly learns the diffusion-based representation from graph structured data and takes it as an effective basis for classification [1]. Specifically, In a convolutional layer, each node in the structure is transformed into a diffusion convolution Z which is combined by its H hop neighbors. And it is defined by an H × F real-valued weight tensor W^c and a nonlinear differentiable function f that computes the activations. The DCNN model is illustrated in Fig. 1. The diffusion-convolutional activation Z_{ijk} for node i, hop j, and feature k of graph is given by

$$Z_{ijk} = f(W^c_{jk} \cdot \sum_{l=0}^{N} P^*_{ijl} X_{lk}),$$ (1)

where P^* is an N × H × N tensor containing the power series of P_j, i.e., $P^* = \{P_0, P_1, ..., P_j, ...P_H\}$, where H is the number of hops. P_j is a degree-normalized transition matrix that gives the probability of jumping from node i to node l in j steps. And we can calculate matrix P_j by adjacency matrix A.

$$P_j = \begin{cases} A, & \text{if } j = 0; \\ AP_{j-1}, & \text{if } j > 0. \end{cases}$$ (2)

The activations can be expressed more concisely using tensor notation as

$$Z = f(W^c \odot P^* X),$$ (3)

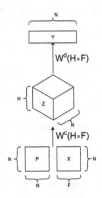

Fig. 1. DCNN model.

where the $\odot$ operator represents element-wise multiplication. After the propagation and activation of the neural network, the probability distribution $\mathbb{P}(Y|X)$ and the final prediction label of $\hat{Y}$ can be obtained by [1]:

$$\hat{Y} = argmax(f(W^d \odot Z)), \tag{4}$$

$$\mathbb{P}(Y|X) = softmax(f(W^d \odot Z)). \tag{5}$$

We present the proposed algorithm of DCNN for social spammer detection in Algorithm 1.

Algorithm 1: Algorithm 1. DCNN Algorithm

   Input: feature information X, social network information R, tag data Y,
           number of hops H, number of features K, Iteration times N
   Output: predict $\hat{Y}$
1   Get A through R;
2   Get P^* according to Eq.(2);
3   Initialization n = 0;
4   **while** *not convergence or n <N* **do**
5   Obtain Z according to Eq.(1) through the convolutional layer;
6   Obtain $\hat{Y}$ and $\mathbb{P}(Y|X)$ according to Eq.(4) and Eq.(5) through the fully connection layer;
7   n = n + 1
8   return $\hat{Y}$;

2.3 DCNN+CR Model

We propose a modified DCNN model named DCNN+CR from two aspects: 1. attenuation of information in diffusion. 2. using social regularization to limit diffusion. In order to prove the validity of each modification point, we have done experiments on each modification point i.e., DCNN+C and DCNN+R.

Attenuation Simulation. Through the introduction above, we know that the core of DCNN model is to obtain diffusion information from other users through the graph structure, and apply them to a convolutional neural network. However, we know that energy will always be decay in the process of transmission for some reason. Therefore, we propose to add a attenuation coefficient w that is a decimal between 0 and 1 to the diffusion information to simulate the information attenuation. Our modified formula is as follows:

$$P_j = \begin{cases} A, & \text{if } j = 0; \\ wAP_{j-1}, & \text{if } j > 0. \end{cases} \tag{6}$$

Regularization of Social Relations. Researchers have used this social relationship to detect spammers, for example, using the social relationship to regularize the decomposition of feature matrix [5,16].

There are four kinds of relationships between legitimate users and spammers as in Fig. 2. Such as, legitimate user → legitimate user(L1 → L2 and L2 → L1) and spammer → spammer(S1 → S2 and S2 → S1) [11]. Many researchers have used the information in the above picture. For example, Y. Zhu et al. in [16], have used legitimate user → legitimate user and spammer → legitimate user relationships in the Fig. 2. And they turned it into a conclusion that legitimate users behaves usually similarly to his friends, in contrast, spammers behave differently from his friends. However, Spammer are very good at hiding and camouflage themselves, for example, they may carefully build their own internal communities to hide themselves and attract legitimate users to follow [3].

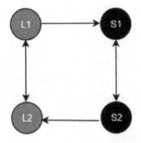

Fig. 2. Social relationships among users.

Therefore, based on the above analysis, we can draw a conclusion: users who follow each other have similar behavior characteristics, while users who only have one edge of following have different behavior characteristics. We can use it to limit the transition matrix. Our modified formula is as follows:

$$\begin{cases} A_{ij} = A_{ji} = 1, & \text{when user i and user j pay attention to each other;} \\ A_{ij} = A_{ji} = 0, & \text{when no connection between user i and user j;} \\ A_{ij} = A_{ji} = -1, & \text{other.} \end{cases} \qquad (7)$$

3 Experiment

3.1 Data Processing

We choose a public and standard dataset, Twitter Social Honeypot Dataset [8] used in [4,5], which provides the ground truth data, i.e., labeling users as spammer or legitimate ones, and part of following relationships. In order to complete the whole following relationships among users, we use the other public dataset the Kwak's dataset [7]. First of all, we remove the non-English tweets and the users who posted less than one tweet. Than, we extract the following relationships. Next, we parse twitter content, extract words with high frequency as features, and named these features as text features. Finally, we extract some features which are used in [9] and named these features as UD_FN_C_H. After processing, the final dataset contains 11130 validate users (7160 legitimate users and 3970 spammers).

3.2   Comparative Baselines and Experimental Results

Comparative Baselines. We use accuracy, precision, recall and F1-score to measure the performance of methods. The comparative baselines are introduces briefly as follows.

- **SMFSR** SMFSR [16] is a classic spammer detection method optimized by social network relationships. And it's different from DCNN+CR in using social network relationships, as we said in the introduction.
- **DCNN** DCNN is the original model of our improved model.
- **DCNN+C** This means that we modify the DCNN only with the attenuation coefficient. We prove that the modification is effective through experiments.
- **DCNN+R** It represents the modify DCNN using social regularization.

Table 1. Social spammer detection results by text features.

Method	Training data one (50%)				Training data two (80%)			
	Accuracy	Precision	Recall	F1-score	Accuracy	Precision	Recall	F1-score
SMFSR	0.6427	0.6456	**0.9930**	0.7825	0.6501	0.6524	**0.9942**	0.7878
DCNN	0.8864	0.9300	0.8975	0.9133	0.8956	0.9362	0.9058	0.9206
DCNN+C	0.8885	0.9325	0.8976	0.9147	0.9019	0.9315	0.9184	0.9248
DCNN+R	0.8862	**0.9379**	0.8918	0.9139	0.9018	**0.9419**	0.9091	0.9251
DCNN+CR	**0.8892**	0.9306	0.9003	**0.9150**	**0.9040**	0.9413	0.9132	**0.9270**

Experimental Result. For the experiments in Table 1, we use only text features. But in Table 2, We use combined features (i.e., combining text and UD_FN_C_H features into a feature matrix as the input of the model). We vary the size of training dataset to observe the performance's trend, where "training data one (50%)" means that we randomly select 50% data from the whole dataset as training set and the remaining 50% data as testing set. All the results listed in table the average of 10 runs of the experiments under different random splits.

From Table 1, we can observe DCNN+CR outperform other baselines on different size of the training data which demonstrate the effectiveness of incorporating network structure information for detecting spammers. Besides, DCNN+C is better than DCNN in accuracy and recall for detection spammer. Moreover, DCNN+R is more effective (on precision) than DCNN when detecting spammers on different size of the training set. As for the F1-score they have a greater advantage over the DCNN. Therefore, the experimental results demonstrate the effectiveness of our proposed model on spammer detection. Compared with SMFSR in Table 1, the precision, F1-score and accuracy of DCNN+CR is greater over 10% than theirs on different size of the training set. DCNN+CR not only takes useful diffusion information into consideration, but also takes advantage of different types of constraints among users to improve the performance significantly.

Comparing the experimental results of the same model in Table 1 and Table 2, we can get the some conclusions. Through DCNN, we can see that the performance about accuracy and F1-score of the DCNN will be reduced by the combined features. SMFSR, DCNN+C and DCNN+R are not robust enough. In 50% of training data sets, the accuracy and F1 score performance of DCNN+C model with text features are better than that of DCNN+C model with combined features, while in 80% of training data sets, the opposite is true. Moreover, DCNN+R and SMFSR are similar to DCNN+C model. In addition, the combined features can improve the performance of DCNN+CR.

Table 2. Social spammer detection results by combined features.

Method	Training data one (50%)				Training data two (80%)			
	Accuracy	Precision	Recall	F1-score	Accuracy	Precision	Recall	F1-score
SMFSR	0.6439	0.6439	**0.9978**	0.7827	0.6473	0.6473	**0.9978**	0.7852
DCNN	0.8742	0.9435	0.8831	0.9091	0.8827	0.9401	0.8984	0.9151
DCNN+C	0.8792	0.9440	0.8894	0.9129	0.9116	0.9403	0.9247	0.9323
DCNN+R	0.8505	**0.9546**	0.8564	0.8972	0.9121	0.9419	0.9232	0.9323
DCNN+CR	**0.8994**	0.9436	0.9035	**0.9231**	**0.9221**	**0.9475**	0.9329	**0.9400**

4   Conclusion

In this paper, we proposed a modified graph convolutional network, namely DCNN+CR for spammer detection. We modified it based on energy decay and social regularization in spammers. Experimental results on a real-world dataset show that DCNN+CR obtains better detection performance than the existing methods at any training ration of data.

References

1. Atwood, J., Towsley, D.: Diffusion-convolutional neural networks. In: Advances in Neural Information Processing Systems 29: Annual Conference on Neural Information Processing Systems 2016, 5–10 December 2016, Barcelona, Spain, pp. 1993–2001 (2016)
2. Chu, Z., Widjaja, I., Wang, H.: Detecting social spam campaigns on twitter. In: Applied Cryptography and Network Security - 10th International Conference, ACNS 2012, Proceedings, Singapore, June 26–29, pp. 455–472 (2012)

3. Ghosh, S., et al.: Understanding and combating link farming in the twitter social network. In: Proceedings of the 21st World Wide Web Conference 2012, WWW 2012, Lyon, France, 16–20 April 2012, pp. 61–70 (2012)

4. Hu, X., Tang, J., Gao, H., Liu, H.: Social spammer detection with sentiment information. In: 2014 IEEE International Conference on Data Mining, ICDM 2014, Shenzhen, China, 14–17 December 2014, pp. 180–189 (2014)

5. Hu, X., Tang, J., Liu, H.: Online social spammer detection. In: Proceedings of the Twenty-Eighth AAAI Conference on Artificial Intelligence, Québec City, Québec, Canada, 27–31 July, pp. 59–65 (2014)

6. Hu, X., Tang, J., Zhang, Y., Liu, H.: Social spammer detection in microblogging. In: IJCAI 2013, Proceedings of the 23rd International Joint Conference on Artificial Intelligence, Beijing, China, 3–9 August 2013, pp. 2633–2639 (2013)

7. Kwak, H., Lee, C., Park, H., Moon, S.B.: What is twitter, a social network or a news media? In: Proceedings of the 19th International Conference on World Wide Web, WWW 2010, Raleigh, North Carolina, USA, 26–30 April 2010, pp. 591–600 (2010)

8. Lee, K., Caverlee, J., Webb, S.: Uncovering social spammers: social honeypots + machine learning. In: Proceeding of the 33rd International ACM SIGIR Conference on Research and Development in Information Retrieval, SIGIR 2010, Geneva, Switzerland, 19–23 July 2010, pp. 435–442 (2010)

9. Lee, K., Eoff, B.D., Caverlee, J.: Seven months with the devils: a long-term study of content polluters on twitter. In: Proceedings of the Fifth International Conference on Weblogs and Social Media, Barcelona, Catalonia, Spain, 17–21 July 2011 (2011)

10. Li, C., Wang, S., He, L., Yu, P.S., Liang, Y., Li, Z.: SSDMV: semi-supervised deep social spammer detection by multi-view data fusion. In: IEEE International Conference on Data Mining, ICDM 2018, Singapore, 17–20 November 2018. pp. 247–256 (2018)

11. Shen, H., Ma, F., Zhang, X., Zong, L., Liu, X., Liang, W.: Discovering social spammers from multiple views. Neurocomputing **225**, 49–57 (2017)

12. Wu, F., Shu, J., Huang, Y., Yuan, Z.: Co-detecting social spammers and spam messages in microblogging via exploiting social contexts. Neurocomputing **201**, 51–65 (2016)

13. Yang, C., Harkreader, R.C., Gu, G.: Empirical evaluation and new design for fighting evolving twitter spammers. IEEE Trans. Inf. Forensics Secur. **8**(8), 1280–1293 (2013)

14. Yang, C., Harkreader, R.C., Zhang, J., Shin, S., Gu, G.: Analyzing spammers' social networks for fun and profit: a case study of cyber criminal ecosystem on twitter. In: Proceedings of the 21st World Wide Web Conference 2012, WWW 2012, Lyon, France, 16–20 April 2012, pp. 71–80 (2012)

15. Zheng, X., Zeng, Z., Chen, Z., Yu, Y., Rong, C.: Detecting spammers on social networks. Neurocomputing **159**, 27–34 (2015)

16. Zhu, Y., Wang, X., Zhong, E., Liu, N.N., Li, H., Yang, Q.: Discovering spammers in social networks. In: Proceedings of the Twenty-Sixth AAAI Conference on Artificial Intelligence, 22–26 July, Toronto, Ontario, Canada (2012)

Dynamics of Online Toxicity in the Asia-Pacific Region

Thomas Marcoux$^{(\boxtimes)}$, Adewale Obadimu, and Nitin Agarwal

University of Arkansas at Little Rock, Little Rock, AR 72204, USA
{txmarcoux,amobadimu,nxagarwal}@ualr.edu

Abstract. The adverse effect of online toxicity on social interactions is enormous. The proliferation of smart devices and mobile applications has further exacerbated these nefarious acts on various social media platforms. Toxic behavior can have a negative impact on a community or dissuade others from joining a community by perceiving it as a hostile or unfriendly environment. Despite the rich research on understanding toxicity on a global scale, there is currently a dearth of systematic research regarding toxicity within the user-generated content on online platforms. To further the understanding of this phenomenon, this paper proposes a precise definition and a case study in an applicable, real world, and large-scale scenario. Using blog posts related to Asia-Pacific political conflicts, we study the toxicity associated with the collected online discourse and pair this analysis with visualizations of the related topic models in order to better understand what type of discourse results in online toxicity. Our analysis shows the potential of the toxicity metric as a tool to understand online events by isolating toxic actors and identifying patterns.

Keywords: Mining social networks · Mining behavioral data · Opinion mining and sentiment analysis

1 Introduction

A toxic message is a rude, disrespectful, or unreasonable message that is likely to result in lower quality of discourse or makes one or more party leave a conversation [1]. Therefore, online toxicity analysis is different from sentiment analysis:

This research is funded in part by the U.S. National Science Foundation (OIA-1920920, IIS-1636933, ACI-1429160, and IIS-1110868), U.S. Office of Naval Research (N00014-10-1-0091, N00014-14-1-0489, N00014-15-P-1187, N00014-16-1-2016, N00014-16-1-2412, N00014-17-1-2605, N00014-17-1-2675, N00014-19-1-2336), U.S. Air Force Research Lab, U.S. Army Research Office (W911NF-16-1-0189), U.S. Defense Advanced Research Projects Agency (W31P4Q-17-C-0059), Arkansas Research Alliance, and the Jerry L. Maulden/Entergy Endowment at the University of Arkansas at Little Rock. Any opinions, findings, and conclusions or recommendations expressed in this material are those of the authors and do not necessarily reflect the views of the funding organizations. The researchers gratefully acknowledge the support.

© Springer Nature Switzerland AG 2020
W. Lu and K. Q. Zhu (Eds.): PAKDD 2020 Workshops, LNAI 12237, pp. 80–87, 2020.
https://doi.org/10.1007/978-3-030-60470-7_9

which is an attempt to assign sentiment scores of positive, neutral, and negative to text data. The toxicity of a message is the multi-dimensional representation of this message through different properties that can be construed as toxic [1]. In this study, we are interested in the willful use of such types of message that aim to decrease the quality of online discussion surrounding specific topics, especially of a political nature, as this could constitute a deliberate effort to muzzle free political discourse. We define this along five different dimensions that contribute to the message's toxicity - as described below [1,2].

1. Identity Attack: This refers to the usage of negative or hateful comments that targets someone because of their identity.
2. Insult: These are comments that are inflammatory or negative towards a person or a group of people.
3. Profanity: It is the usage of swear words, curse words, or other obscene or profane language.
4. Threat: This type of comment describes an intention to inflict pain, injury, or violence against an individual or group.
5. Sexually Explicit: These are comments that contains references to sexual acts, body parts, or other lewd content.

2   Literature Review

This study is two-fold. We combine and expand the extensive research done on toxicity with the topic modeling technique in order to find meaningful insight.

2.1   Online Toxicity

Extant literature on this topic suggests that toxicity, in its multiple forms defined above, will tend to generally lower the quality of a conversation in a community [1,2]. Research by [3] shows that the pervasiveness of toxicity in a community can dissuade other people from joining the community due to the perception of the community as being hostile or unfriendly. This is especially true in the context of political discussion and other hot-button issues. Such behaviors becoming commonplace could result in damage to the availability of democratic speech [3].

There are multiple explanations as to why online users may engage in toxic behavior. Some studies have shown that online users tend to participate in toxic behaviors out of boredom [4], others participate to vent [4], or simply to have fun [5]. Other studies suggest that toxic users have unique personality traits and motivation [6], while some others note that, given the right conditions, anyone can exhibit toxic tendencies [7]. [8] shows that toxic users become worse over time. As their comments become more toxic, they are more likely to become intolerant of the community.

The major challenge in understanding toxic behavior is maintaining the balance of freedom of expression and curtailing harmful content. There exist different opinion mining techniques, such as sentiment analysis, which outputs a positive or negative score for a given comment. We choose to use toxicity scores as the chosen metric because toxicity analysis has proven to be more apt at detecting toxic intent behind more subtle, complex, or culturally-dependent speech that cannot be labelled as toxic through simple sentiment analysis alone [1].

2.2   Applying Topic Modeling to Social Media Data

In natural language processing, topic modeling is a statistical model that allows for the discovery of abstract "topics" within a corpus of documents [9]. The corpus is divided into n "topics", each defined as a set of words that best represent one underlying theme of the corpus. The technique we use to model the topics in this paper is Latent Dirichlet Allocation (LDA) [10]. This allows us to observe the topic distribution for each document in our corpus, and the word distribution for each or those topics. Knowing these metrics is necessary for the visualization of topic streams. This is described in further detail in the methodology section. One point of note is that LDA is not infallible and can produce flawed topics or "junk topics" catching miscellaneous words (as seen in our Topics Table - Table 3 - where Topic 8 captured non-speech, scrap from the data collection process).

3   Methodology

In order to gather the needed data, we use the Blogtrackers tool [11,12]. We selected 24 blogs (see Table 1) with content published between 1990 and 2019 and performed manual relevance assessment to determine if they are contextually relevant to the Asia-Pacific (APAC) region for a total of 101,017 posts.

Table 1. Blogs crawled and number of posts.

Blog name	Posts collected	Blog name	Posts collected
dissidentvoice.org	18439	buzz.definitelyfilipino.net	1759
www.counterpunch.org	11765	orientalreview.org	1362
countercurrents.org	9556	www.desmog.co.uk	1293
www.wsws.org	8885	www.geopolitica.ru	1046
www.desmogblog.com	8145	grain.org	1002
www.strategic-culture.org	7315	anddestroyer.blogspot.com	797
www.greenpeace.org	6583	www.asia-pacificresearch.com	673
journal-neo.org	6367	pinoytrending.altervista.org	575
consortiumnews.com	5781	www.zoominkorea.org	381
washingtonsblog.com	4903	ndiagminfo.org	194
gmwatch.org	2038	www.thenewatlas.org	193
www.asianpolicy.press	1959	muslimcyberarmy.blogspot.com	6

For the crawling process, we leveraged the Web Content Extractor (WCE) software [13] to extract the blog post along with its title, the publication date, the name of the blogger, the post geo-location, and the language of the post. We used this tool to train a crawler that extracts data from each individual blog site. This automated crawling process does not always output clean data. The output is sometimes riddled with some noise. Therefore the data needs to be cleaned before analysis is performed. We follow a three step cleaning process.

The second step is to clean the dataset. First we clean from the tool itself (WCE) by purging empty fields and advertisement URLs. Then the data is stored into a temporary database. In step 2 we clean the data through SQL queries. We only select and save validated and verified data from the temporary database. Finally, we use scripts for data processing, metadata extraction, and data standardization. We exclude any possible noise and standardize attributes like date of posting.

4   Results and Discussion

One essential aspect of the analysis is to compute a toxic score for each blog post or comment. To achieve this, we leveraged a classification tool developed by Google's Project Jigsaw and "Counter Abuse Technology" teams: Perspective API [2]. This allows us to assign a score on the perceived toxicity of a given blog post or comment. This model uses a Convolutional Neural Network (CNN) trained with word-vector inputs to determine whether a blog post or comment could be perceived as "toxic" to the community. Specifically, the tool analyzes the five (5) frequently recurring toxic dimensions: threat, insult, profanity, sexually explicit, and identity-based attack. The API returns a probability score between 0 and 1, with higher values indicating a greater likelihood that the content should be given an overall label of "toxic", and a greater likelihood of the content being assigned specific dimension labels. Below are the research questions we strive to answer using these tools.

4.1   Toxicity Analysis

What is the trend of average toxicity over a period of time?
Here, we sought to understand how a shift in the context of the collective blog posts influences the temporal pattern of toxicity in our dataset. We computed the average toxicity of the cumulative blog posts for each year and plotted it against the yearly interval. We observed some temporal patterns in the plot. For instance, despite having similar total number of blog posts, there was a slight decrease in average toxicity from 1990 to 1996. We also observed a steady rise in the annual average toxicity from 2004 to 2009 (see Fig. 1).

What is the influence of a post's content on the average toxicity of the blogs?
Term frequency analysis revealed that some of the prominent keywords used in the blog posts in 1996 are related to "rights", "knowledge", "property", and "farming" – which makes the average toxicity for this year to be understandably low. This implies that the content of posts in this year is innocuous. However, from 2004 to 2007, political blogs are rising. As a result we witness a drastic rise

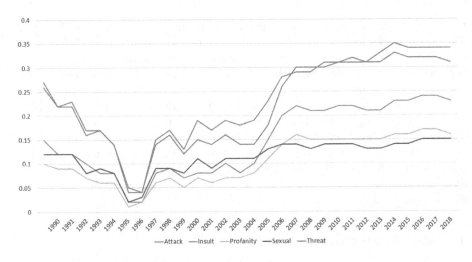

Fig. 1. Average toxicity for all toxic dimensions.

in the average toxicity for these periods. Term frequency analysis of the blog posts within these periods shows a gradual progression from political discussion to a high frequency in the usage of words that are related to "war", "military" and "nuclear".

Does the length of blog posts influence the average toxicity?
We hypothesize that the length of a blog post does not have a direct effect on the average toxicity of the blog posts. We observe that despite the decline in the average toxicity from 1994 to 1996, the amount of blog posts is relatively steady. Further statistical tests are needed to further corroborate this claim.

Is toxicity a better indicator than existing measures?
Currently, there are several existing text-based indicators that have been used to understand the orientation of the blogosphere. For instance, sentiment analysis has been used extensively in the literature to understand the opinion and tonality of blogs. In the same vein, influence analysis has been used to understand the driving force in the blogosphere. Here, we hypothesize that toxicity should be an additional measure that can be used to further understand the orientation of the blogosphere. Figure 2 shows a comparison of the average sentiment and average toxicity in the dataset. Despite having identical trends, we observed that average toxicity was able to capture the variability better than average sentiment.

4.2   Topic Modeling

In order to provide context for toxicity scores and also explore the topics used within the scope of Asia-Pacific blogs, we employed different statistical techniques and visualization methods. One such method is the visualization of topic streams (Fig. 3). Topic streams are a visual representation of the frequency of each topic against a chronological axis. Table 2 provides an example of the

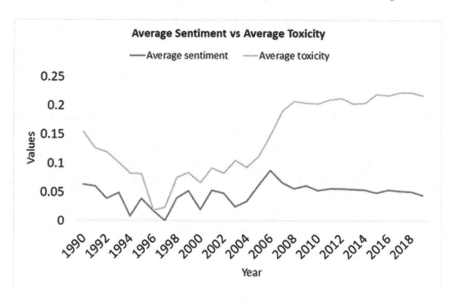

Fig. 2. Average sentiment and average toxicity per year.

results we obtained from using the Python module Gensim [14] to apply Latent Dirichlet Allocation. Gensim takes the entire corpus of blog posts as an input - along with a set of stopwords, words to be ignored by the model - and outputs the topic distribution for a number K of topics (set to k = 10 in this case). In Table 2 we see the result for one blog post m. The range of the distribution is the probability from 0 to 1 corresponding to the likelihood that a particular document will be categorized as falling into a particular "topic". For this particular blog post, "Topic 6" shows the highest probability score, which means that this particular post has an overall topic that is most likely illustrated by the characteristics that the LDA model has designated as "Topic 6".

Table 2. Sample of topic distribution for a post m.

Post	Topic 0	Topic 1	Topic 2	Topic 3	Topic 4	Topic 5	Topic 6	Topic 7	Topic 8	Topic 9
m	0.04964	0.00020	0.00020	0.16115	0.01770	0.00717	0.41380	0.05258	0.23334	0.0641

In Table 3, we reference the main ideas behind the 10 topics we generated using Gensim [14]. We use topic 8 as a "junk" topic. That is, a topic that will serve as a "catch-all" for leftover text from the data collection process such as improper web code. We have removed Topic 8 from all visualizations as it offers no insight, along with Topic 5 because it appears uniformly in most blog posts.

One way to observe this data is through a representation of each topic distribution against each blog post (represented by the year of publication). While this

Table 3. Description for each topic.

Topic 0	Topic 1	Topic 2	Topic 3	Topic 4	Topic 5	Topic 6	Topic 7	Topic 8	Topic 9
News	War	Health	Society	Asia	General	Economy	America	Junk	Law

method has the advantage of accurately representing the size of the data set for each year, we choose to use different visualizations to focus on the chronological distribution of the data rather than focus on the yearly volume of data.

In fact, when looking at values averaged per year, early blogs disappear. For this reason, we averaged the values of the topic distributions for each day and applied a moving average with a window size of 50 to smooth the curve, see Fig. 3. Area charts are the preferred visualization tool as they allow for an easier visualization and low value overlaps become hidden - as opposed to a line chart view which becomes confusing due to the multitude of topics.

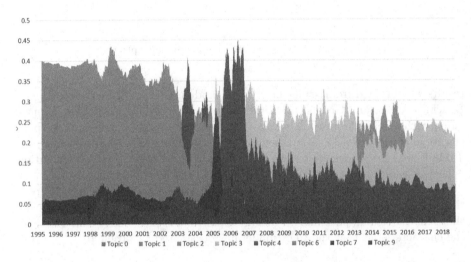

Fig. 3. Average topic distribution by day, smoothed by a moving average of k = 50.

One of the main points of interest of the topic streams shown in Fig. 3 is the obvious divide between pre and post 2005. We can see that, pre-2005, online discourse is dominated by Topic 6, which seems centered around economical questions. Interestingly, economical questions as represented by Topic 6 plummeted around 2005 and are only slowly starting to recover. At the same time, we notice a rise in Topics 3 and 4 - topics associated with Chinese and Korean issues, as well as social and political matters. Conversely, it would seem that online discourse in the Asia-Pacific region has pivoted from purely economical concerns to a more politically and internationally oriented online climate.

While any number of reasons can explain online toxicity, the data we found in this study seem to suggest an overarching theme. The analysis of toxicity scores sees a huge rise around 2005. When put next to the dynamics of topic streams, this seems to coincide with a move of the online discussion towards a more outward look at politics (Chinese and Korean questions, social issues, etc.).

5   Conclusion and Future Work

This analysis highlights the potential of the toxicity metric as an indicator for understanding the shift in the orientation of the blogosphere. One way to improve this study will be to use different statistical techniques to make up for the disparity in chronological distribution. We also plan to conduct further analysis to understand how this measure can be used to extensively understand blogosphere dynamics. For instance, we plan to identify who the most toxic bloggers are and what are the patterns of their posts. Another question we plan to answer is whether the most influential bloggers are also the most toxic on average.

The data and methods used in this study resulted in models that match the ground truth around APAC events and encourage a more zoomed-in study of different scenarios such as the Honk-Kong crisis, which could pave the way for a predictive model of real world events from online social dynamics.

References

1. Obadimu, A., Mead, E., Hussain, M., Agarwal, N.: Identifying toxicity within YouTube video comment text data (2019). https://doi.org/10.13140/RG.2.2. 15254.19522
2. Google's perspective API. https://www.perspectiveapi.com/#/home
3. Lee, S.-H., Kim, H.-W.: Why people post benevolent and malicious comments online. Commun. ACM **58**, 74–79 (2015). https://doi.org/10.1145/2739042
4. Varjas, K., Talley, J., Meyers, J., Parris, L., Cutts, H.: High school students' perceptions of motivations for cyberbullying: an exploratory study. West J. Emerg. Med. **11**(3), 269–273 (2010)
5. Shachaf, P., Hara, N.: Beyond vandalism: Wikipedia trolls. J. Inf. Sci. **36**(3), 357–370 (2010). https://doi.org/10.1177/0165551510365390
6. Suler, J.: The online disinhibition effect. **7**. https://doi.org/10.1089/1094931041291295
7. Cheng, J., Bernstein, M., Danescu-Niculescu-Mizil, C., Leskovec, J.: Anyone can become a troll: causes of trolling behavior in online discussions. https://doi.org/10.1145/2998181.2998213
8. Cheng, J., Danescu-Niculescu-Mizil, C., Leskovec, J.: Antisocial behavior in online discussion communities. 10 (2015)
9. Blei, D.M., Lafferty, J.D.: Topic models (2009)
10. Blei, D.M., Ng, A.Y., Jordan, M.I.: Latent Dirichlet allocation. J. Mach. Learn. Res. **3**, 993–1022 (2003)
11. Hussain, M., Obadimu, A., Bandeli, K.K., Nooman, M., Al-khateeb, S., Agarwal, N.: A framework for blog data collection: challenges and opportunities (2017)
12. Obadimu, A., Hussain, M., Agarwal, N.: Blog data analytics using blogtrackers (2019). https://doi.org/10.1145/3341161.3343707
13. Web content extractor. http://www.newprosoft.com/web-content-extractor.htm
14. Python - Gensim. https://radimrehurek.com/gensim/

Workshop on Learning Data Representation for Clustering (LDRC 2020)

Predictive K-means with Local Models

Vincent Lemaire[1(✉)], Oumaima Alaoui Ismaili[1], Antoine Cornuéjols[2], and Dominique Gay[3]

[1] Orange Labs, Lannion, France
vincent.lemaire@orange.com
[2] AgroParisTech, Université Paris-Saclay, Paris, France
[3] LIM-EA2525, Université de La Réunion, Saint-Denis, France

Abstract. Supervised classification can be effective for prediction but sometimes weak on interpretability or explainability (XAI). Clustering, on the other hand, tends to isolate categories or profiles that can be meaningful but there is no guarantee that they are useful for labels prediction. Predictive clustering seeks to obtain the best of the two worlds. Starting from labeled data, it looks for clusters that are as pure as possible with regards to the class labels. One technique consists in tweaking a clustering algorithm so that data points sharing the same label tend to aggregate together. With distance-based algorithms, such as k-means, a solution is to modify the distance used by the algorithm so that it incorporates information about the labels of the data points. In this paper, we propose another method which relies on a change of representation guided by class densities and then carries out clustering in this new representation space. We present two new algorithms using this technique and show on a variety of data sets that they are competitive for prediction performance with pure supervised classifiers while offering interpretability of the clusters discovered.

1 Introduction

While the power of predictive classifiers can sometimes be awesome on given learning tasks, their actual usability might be severely limited by the lack of interpretability of the hypothesis learned. The opacity of many powerful supervised learning algorithms has indeed become a major issue in recent years. This is why, in addition to good predictive performance as standard goal, many learning methods have been devised to provide readable decision rules [3], degrees of beliefs, or other easy to interpret visualizations. This paper presents a predictive technique which promotes interpretability, explainability as well, in its core design.

The idea is to combine the predictive power brought by supervised learning with the interpretability that can come from the descriptions of categories, profiles, and discovered using unsupervised clustering. The resulting family of techniques is variously called *supervised clustering* or *predictive clustering*. In the literature, there are two categories of predictive clustering. The first family of algorithms aims at optimizing the trade-off between description and prediction, i.e., aiming at detecting sub-groups in each target class. By contrast, the

W. Lu and K. Q. Zhu (Eds.): PAKDD 2020 Workshops, LNAI 12237, pp. 91–103, 2020.
https://doi.org/10.1007/978-3-030-60470-7_10

algorithms in the second category favor the prediction performance over the discovery of all underlying clusters, still using clusters as the basis of the decision function. The hope is that the predictive performance of predictive clustering methods can approximate the performances of supervised classifiers while their descriptive capability remains close to the one of pure clustering algorithms.

Several predictive clustering algorithms have been presented over the years, for instance [1,4,10,11,23]. However, the majority of these algorithms require (*i*) a considerable execution time, and (*ii*) that numerous user parameters be set. In addition, some algorithms are very sensitive to the presence of noisy data and consequently their outputs are not easily interpretable (see [5] for a survey). This paper presents a new predictive clustering algorithm. The underlying idea is to use any existing distance-based clustering algorithms, e.g., k-means, but on a redescription space where the target class is integrated. The resulting algorithm has several desirable properties: there are few parameters to set, its computational complexity is almost linear in m, the number of instances, it is robust to noise, its predictive performance is comparable to the one obtained with classical supervised classification techniques and it tends to produce groups of data that are easy to interpret for the experts.

The remainder of this paper is organized as follows: Sect. 2 introduces the basis of the new algorithm, the computation of the clusters, the initialization step and the classification that is realized within each cluster. The main computation steps of the resulting predictive clustering algorithms are described in Algorithm 1. We then report experiments that deal with the predictive performance in Sects. 3. We focus on the supervised classification performance to assess if predictive clustering could reach the performances of algorithms dedicated to supervised classification. Our algorithm is compared using a variety of data sets with powerful supervised classification algorithms in order to assess its value as a predictive technique. And an analysis of the results is carried out. Conclusion and perspectives are discussed in Sect. 4.

2   Turning the K-means Algorithm Predictive

The k-means algorithm is one of the simplest yet most commonly used clustering algorithms. It seeks to partition m instances $(X_1, \ldots X_m)$ into K groups $(B_1, \ldots, B_K)$ so that instances which are close are assigned to the same cluster while clusters are as dissimilar as possible. The objective function can be defined as:

$$\mathcal{G} = \underset{B_i}{\text{Argmin}} \sum_{i=1}^{K} \sum_{X_j \in B_i} \|X_j - \mu_i\|^2 \tag{1}$$

where μ_i are the centers of clusters B_i and we consider the Euclidean distance.

Predictive clustering adds the constrain of maximizing clusters purity (i.e. instances in a cluster should share the same label). In addition, the goal is to provide results that are easy to interpret by the end users. The objective function of Eq. (1) needs to be modified accordingly.

One approach is to modify the distance used in conventional clustering algorithm in order to incorporate information about the class of the instances. This modified distance should make points differently labelled appear as more distant than in the original input space. Rather than modifying the distance, one can instead alter the input space. This is the approach taken in this paper, where the input space is partitioned according to class probabilities prior to the clustering step, thus favoring clusters of high purity. Besides the introduction of a technique for computing a new feature space, we propose as well an adapted initialization method for the modified k-means algorithm. We also show the advantage of using a specific classification method within each discovered cluster in order to improve the classification performance. The main steps of the resulting algorithm are described in Algorithm 1. In the remaining of this Sect. 2 we show how each step of the usual K-means is modified to yield a predictive clustering algorithm.

Algorithm 1. Predictive K-means algorithm

Input:
- D: a data set which contains m instances. Each one (X_i) $i \in \{1, \ldots, m\}$ is described by d descriptive features and a label $C_i \in \{1, \ldots, J\}$.
- K: number of clusters.

Start:
1) Supervised preprocessing of data to represent each X_i as $\widehat{X}_i$ in a new feature space $\Phi(\mathcal{X})$.
2) Supervised initialization of centers.

repeat
    3) *Assignment:* generate a new partition by assigning each instance $\widehat{X}_i$ to the nearest cluster.
    4) Representation: calculate the centers of the new partition.

until the convergence of the algorithm

5) Assignment classes to the obtained clusters:
        - *method 1*: majority vote.
        - *method 2*: local models.

6) Prediction the class of the new instances in the deployment phase:
        $\rightarrow$ *the closest cluster class (if method 1 is used).*
        $\rightarrow$ *local models (if method 2 is used).*

End

A Modified Input Space for Predictive Clustering - The principle of the proposed approach is to partition the input space according to the class probabilities $P(C_j|X)$. More precisely, let the input space $\mathcal{X}$ be of dimension d, with numerical descriptors as well as categorical ones. An example $X_i \in \mathcal{X}$ $(X_i = [X_i^{(1)}, \ldots, X_i^{(d)}]^\top)$ will be described in the new feature space $\Phi(\mathcal{X})$ by $d \times J$ components, with J being the number of classes. Each component $X_i^{(n)}$ of $X_i \in \mathcal{X}$ will give J components $X_i^{(n,j)}$, for $j \in \{1, \ldots, J\}$, of the new description $\widehat{X}_i$ in $\Phi(\mathcal{X})$, where $\widehat{X}_i^{(n,j)} = \log P(X^{(n)} = X_i^{(n)}|C_j)$, i.e., the log-likelihood values.

Therefore, an example X is redescribed according to the (log)-probabilities of observing the values of original input variables given each of the J possible classes (see Fig. 1). Below, we describe a method for computing these values. But first, we analyze one property of this redescription in $\Phi(\mathcal{X})$ and the distance this can provide.

$\mathcal{X}$	$X^{(1)}$	...	$X^{(d)}$	Y
X_1		...		
...		...		
X_m		...		

$\overset{\Phi}{\Longrightarrow}$

$\Phi(\mathcal{X})$	$X^{(1,1)}$	...	$X^{(1,J)}$	...	$X^{(d,1)}$	...	$X^{(d,J)}$	Y
X_1				...				
...				...				
X_m				...				

Fig. 1. Φ redescription scheme from d variables to $d \times J$ variables, with log-likelihood values: $\log P(X^{(n)}|C_j)$

Property of the Modified Distance - Let us denote $dist_B^p$ the new distance defined over $\Phi(\mathcal{X})$. For the two recoded instances $\hat{X}_1$ and $\hat{X}_2 \in \mathbb{R}^{d \times J}$, the formula of $dist_B^p$ is (in the following we omit $\hat{X} = \hat{X}_i$ in the probability terms for notation simplification):

$$dist_B^p(\hat{X}_1, \hat{X}_2) = \sum_{j=1}^{J} \| \log(P(\hat{X}_1|C_j)) - \log(P(\hat{X}_2|C_j)) \|_p \qquad (2)$$

where $\| \cdot \|_p$ is a Minkowski distance. Let us denote now, $\Delta^p(\hat{X}_1, \hat{X}_2)$ the distance between the (log)-posterior probabilities of two instances $\hat{X}_1$ and $\hat{X}_2$. The formula of this distance as follow:

$$\Delta^p(\hat{X}_1, \hat{X}_2) = \sum_{j=1}^{J} \| \log(P(C_j|\hat{X}_1)) - \log(P(C_j|\hat{X}_2)) \|_p \qquad (3)$$

where $\forall i \in \{1, \dots, m\}$, $P(C_j|\hat{X}_i) = \frac{P(C_j) \prod_{n=1}^{d} P(X_i^{(n)}|C_j)}{P(\hat{X}_i)}$ (using the hypothesis of features independence conditionally to the target class). From the distance given in Eq. 3, we find the following inequality:

$$\Delta^p(\hat{X}_1, \hat{X}_2) \leq \{dist_B^p(\hat{X}_1, \hat{X}_2) + J \| \log(P(\hat{X}_2)) - \log(P(\hat{X}_1)) \| \} \qquad (4)$$

Proof.

$$\Delta^p = \sum_{j=1}^{J} \| \log(P(C_j|\hat{X}_1)) - \log(P(C_j|\hat{X}_2)) \|_p$$

$$= \sum_{j=1}^{J} \| \log(\frac{P(\hat{X}_1|C_j)P(C_j)}{P(\hat{X}_1)}) - \log(\frac{P(\hat{X}_2|C_j)P(C_j)}{P(\hat{X}_2)}) \|_p$$

$$= \sum_{j=1}^{J} \| \log(P(\hat{X}_1|C_j)) - \log(P(\hat{X}_1)) - \log(P(\hat{X}_2|C_j)) + \log(P(\hat{X}_2)) \|_p$$

$$\leq \sum_{j=1}^{J} [A + B]$$

with $\Delta^p = \Delta^p(\hat{X}_1, \hat{X}_2)$ and $A = \| \log(P(\hat{X}_1|C_j)) - \log(P(\hat{X}_2|C_j)) \|_p$
and $B = \| \log(P(\hat{X}_2)) - \log(P(\hat{X}_1)) \|_p$
then $\Delta^p \leq dist^p_B(\hat{X}_1, \hat{X}_2) + J \| \log(P(\hat{X}_2)) - \log(P(\hat{X}_1)) \|_p$.

This above inequality expresses that two instances that are close in terms of distance $dist^p_B$ will also be close in terms of their probabilities of belonging to the same class. Note that the distance presented above can be integrated into any distance-based clustering algorithms.

Building the Log-Likelihood Redescription $\Phi(\mathcal{X})$ - Many methods can estimate the new descriptors $X_i^{(n,j)} = \log P(X^{(n)} = X_i^{(n)}|C_j)$ from a set of examples. In our work, we use a supervised discretization method for numerical attributes and a supervised grouping values for categorical attributes to obtain respectively intervals and group values in which $P(X^{(n)} = X_i^{(n)}|C_j)$ could be measured. The used supervised discretization method is described in [8] and the grouping method in [7]. The two methods have been compared with extensive experiments to corresponding state of the art algorithms. These methods computes univariate partitions of the input space using supervised information. It determines the partition of the input space to optimize the prediction of the labels of the examples given the intervals in which they fall using the computed partition. The method finds the best partition (number of intervals and thresholds) using a Bayes estimate. An additional bonus of the method is that outliers are automatically eliminated and missing values can be imputed.

Initialisation of Centers - Because clustering is a NP-hard problem, heuristics are needed to solve it, and the search procedure is often iterative, starting from an initialized set of prototypes. One foremost example of many such distance-based methods is the k-means algorithm. It is known that the initialization step can have a significant impact both on the number of iterations and, more importantly, on the results which correspond to local minima of the optimization criterion (such as Eq. 1 in [20]). However, by contrast to the classical clustering methods, predictive clustering can use supervised information for the choice of the initial prototypes. In this study, we chose to use the K++R method. Described in [17], it follows an "exploit and explore" strategy where the class labels are first exploited before the input distribution is used for exploration in order to get the apparent best initial centers. The main idea of this method is to dedicate one center per class (comparable to a *"Rocchio"* [19] solution). Each center is defined as the average vector of instances which have the same class label. If the predefined number of clusters (K) exceeds the number of classes (J), the initialization continues using the K-means++ algorithm [2] for the $K - J$ remaining centers in such a way to add diversity. This method can only be used when $K \geq J$, but this is fine since in the context of supervised clustering[1] we do not look for clusters where $K < J$. The complexity of this scheme is $\mathcal{O}(m + (K - J)m) < \mathcal{O}(mK)$, where m is the number of examples. When $K = J$, this method is deterministic.

---

[1] In the context of supervised clustering, it does not make sense to cluster instances in K clusters where $K < J$.

Instance Assignment and Centers Update - Considering the Euclidean distance and the original K-means procedure for updating centers, at each iteration, each instance is assigned to the nearest cluster (j) using the ℓ_2 metric $(p = 2)$ in the redescription space $\Phi(\mathcal{X})$. The K centers are then updated according to the K-Means procedure. This choice of distance (Euclidean) in the adopted k-means strategy could have an influence on the (predictive) relevance of the clusters but has not been studied in this paper.

Label Prediction in Predictive Clustering - Unlike classical clustering which aims only at providing a description of the available data, predictive clustering can also be used in order to make prediction about new incoming examples that are unlabeled.

The commonest method used for prediction in predictive K-means is the majority vote. A new example is first assigned to the cluster of its nearest prototype, and the predicted label is the one shared by the majority of the examples of this cluster. This method is not optimal. Let us call P_M the frequency of the majority class in a given cluster. The true probability μ of this class obeys the Hoeffding inequality: $P\big(|P_M - \mu| \geq \varepsilon\big) \leq 2\exp(-2\,m_k\,\varepsilon^2)$ with m_k the number of instances assigned to the cluster k. If there are only 2 classes, the error rate is $1 - \mu$ if P_M and μ both are > 0.5. But the error rate can even exceed 0.5 if $P_M > 0.5$ while actually $\mu < 0.5$. The analysis is more complex in case of more than two classes. It is not the object of this paper to investigate this further. But it is apparent that the majority rule can often be improved upon, as is the case in classical supervised learning.

Another evidence of the limits of the majority rule is provided by the examination of the ROC curve [12]. Using the majority vote to assign classes for the discovered clusters generates a ROC curve where instances are ranked depending on the clusters. Consequently, the ROC curve presents a sequence of steps. The area under the ROC curve is therefore suboptimal compared to a ROC curve that is obtained from a more refined ranking of the examples, e.g., when class probabilities are dependent upon each example, rather than groups of examples.

One way to overcome these limits is to use local prediction models in each cluster, hoping to get better prediction rules than the majority one. However, it is necessary that these local models: 1) can be trained with few instances, 2) do not overfit, 3) ideally, would not imply any user parameters to avoid the need for local cross-validation, 4) have a linear algorithmic complexity $O(m)$ in a learning phase, where m is the number of examples, 5) are not used in the case where the information is insufficient and the majority rule is the best model we can hope for, 6) keep (or even improve) the initial interpretation qualities of the global model. Regarding item (1), a large study has been conducted in [22] in order to test the prediction performances in function of the number of training instance of the most commonly classifiers. One prominent finding was that the Naive Bayes (NB) classifier often reaches good prediction performances using only few examples (Bouchard & Triggs's study [6] confirms this result). This fact remains valid even when features receive weights (e.g., Averaging Naive Bayes (ANB) and Selective Naive Bayes (SNB) [16]). We defer discussion of the other items to the Sect. 3 on experimental results.

In our experiments, we used the following procedure to label each incoming data point X: i) X is redescribed in the space $\Phi(\mathcal{X})$ using the method described in Sect. 2, ii) X is assigned to the cluster k corresponding to the nearest center iii) the local model, l, in the corresponding cluster is used to predict the class of X (and the probability memberships) if a local model exits: $P(j|X) = \text{argmax}_{1 \leq j \leq J}(P_{SNB_l}(C_j|X))$ otherwise the majority vote is used (Note: $P_{SNB_l}(C_j|X))$ is described in the next section).

3   Comparison with Supervised Algorithm

3.1   The Chosen Set of Classifiers

To test the ability of our algorithm to exhibit high predictive performance while at the same time being able to uncover interesting clusters in the different data sets, we have compared it with three powerful classifiers (in the spirit of, or close to our algorithm) from the state of the art: Logistic Model Tree (LMT) [15], Naives Bayes Tree (NBT) [14] and Selective Naive Bayes (SNB) [9]. This section briefly described these classifiers.

- **Logistic Model Tree (LMT)** [15] combines logistic regression and decision trees. It seeks to improve the performance of decision trees. Instead of associating each leaf of the tree to a single label and a single probability vector (piecewise constant model), a logistic regression model is trained on the instances assigned to each leaf to estimate an appropriate vector of probabilities for each test instance (piecewise linear regression model). The logit-Boost algorithm is used to fit a logistic regression model at each node, and then it is partitioned using information gain as a function of impurity.
- **Naives Bayes Tree (NBT)** [14] is a hybrid algorithm, which deploys a naive Bayes classifier on each leaf of the built decision tree. NBT is a classifier which has often exhibited good performance compared to the standard decision trees and naive Bayes classifier.
- **Selective Naive Bayes (SNB)** is a variant of NB. One way to average a large number of selective naive Bayes classifiers obtained with different subsets of features is to use one model only, but with features weighting [9]. The Bayes formula under the hypothesis of features independence conditionally to classes becomes: $P(j|X) = \frac{P(j) \prod_f P(X^f|j)^{W_f}}{\sum_{j=1}^{K}[P(j) \prod_f P(X^f|j)^{W_f}]}$, where W_f represents the weight of the feature f, X^f is component f of X, j is the class labels. The predicted class j is the one that maximizes the conditional probability $P(j|X)$. The probabilities $P(X_i|j)$ can be estimated by interval using a discretization for continuous features. For categorical features, this estimation can be done if the feature has few different modalities. Otherwise, grouping into modalities is used. The resulting algorithm proves to be quite efficient on many real data sets [13].

- **Predictive K-Means** (PKM$_{MV}$, PKM$_{SNB}$): (i) PKM$_{VM}$ corresponds to the Predictive K-Means described in Algorithm 1 where prediction is done according to the Majority Vote; (ii) PKM$_{SNB}$ corresponds to the Predictive K-Means described in Algorithm 1 where prediction is done according to a local classification model.
- **Unsupervised K-Means** (KM$_{MV}$) is the usual unsupervised K-Means with prediction done using the Majority Vote in each cluster. This classifier is given for comparison as a baseline method. The pre-processing is not supervised and the initialization used is k-means++ [2] (in this case since the initialization is not deterministic we run k-means 25 times and we keep the best initialization according to the Mean Squared Error). Among the existing unsupervised pre-processing approaches [21], depending on the nature of the features, continuous or categorical, we used:
  - for Numerical attribute: Rank Normalization (RN). The purpose of rank normalization is to rank continuous feature values and then scale the feature into $[0, 1]$. The different steps of this approach are: (i) rank feature values u from lowest to highest values and then divide the resulting vector into H intervals, where H is the number of intervals, (ii) assign for each interval a label $r \in \{1, ..., H\}$ in increasing order, (iii) if X_{iu} belongs to the interval r, then $X'_{iu} = \frac{r}{H}$. In our experiments, we use $H = 100$.
  - for Categorical attribute: we chose to use a Basic Grouping Approach (BGB). It aims at transforming feature values into a vector of Boolean values. The different steps of this approach are: (i) group feature values into g groups with as equal frequencies as possible, where g is a parameter given by the user, (ii) assign for each group a label $r \in \{1, ..., g\}$, (iii) use a full disjunctive coding. In our experiments, we use $g = 10$.

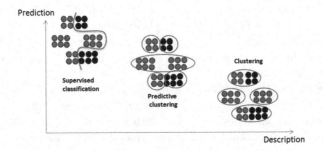

Fig. 2. Differences between the three types of "classification"

In the Fig. 2 we suggest a two axis figure to situate the algorithms described above: a vertical axis for their ability to describe (explain) the data (from low to high) and horizontal axis for their ability to predict the labels (from low to high). In this case the selected classifiers exemplify various trade-offs between prediction performance and explanatory power: (i) KM$_{MV}$ more dedicated to description would appear in the bottom right corner; (ii) LMT, NBT and SNB dedicated to prediction would go on the top left corner; and (iii) PKM$_{VM}$, PKM$_{SNB}$ would lie in between. Ideally, our algorithm, PKM$_{SNB}$ should place itself on the top right quadrant of this kind of figure with both good prediction and description performance.

Note that in the reported experiments, $K = J$ (*i.e*, number of clusters = number of classes). This choice which biases the algorithm to find one cluster per class, is detrimental for predictive clustering, thus setting a lower bound on the performance that can be expected of such an approach.

3.2   Experimental Protocol

The comparison of the algorithms have been performed on 8 different datasets of the UCI repository [18]. These datasets were chosen for their diversity in terms of classes, features (categorical and numerical) and instances number (see Table 1).

Table 1. The used datasets, V_n: numerical features, V_c: categorical features.

Datasets	Instances	#V_n	#V_c	# Classes	Datasets	Instances	#V_n	#V_c	# Classes
Glass	214	10	0	6	Waveform	5000	40	0	3
Pima	768	8	0	2	Mushroom	8416	0	22	2
Vehicle	846	18	0	4	Pendigits	10992	16	0	10
Segmentation	2310	19	0	7	Adult	48842	7	8	2

Evaluation of the Performance: In order to compare the performance of the algorithms presented above, the same folds in the train/test have been used. The results presented in Sect. 3.3 are those obtained in the test phase using a 10×10 folds cross validation (stratified). The predictive performance of the algorithms are evaluated using the AUC (area under the ROC's curve). It is computed as follows: AUC $= \sum_i^C P(C_i)$AUC(C_i), where AUC(i) denotes the AUC's value in the class i against all the others classes and $P(Ci)$ denotes the prior on the class i (the elements frequency in the class i). AUC(i) is calculated using the probability vector $P(C_i|X)$ $\forall i$.

3.3   Results

Performance Evaluation: Table 2 presents the predictive performance of LMT, NBT, SNB, our algorithm PKM$_{MV}$, PKM$_{SNB}$ and the baseline KM$_{MV}$ using the ACC (accuracy) and the AUC criteria (presented as a %). These results show the very good prediction performance of the PKM$_{SNB}$ algorithm. Its performance is indeed comparable to those of LMT and SNB which are the strongest ones. In addition, the use of local classifiers (algorithm PKM$_{SNB}$) provides a clear advantage over the use of the majority vote in each cluster as done in PKM$_{MV}$. Surprisingly, PKM$_{SNB}$ exhibits

Table 2. Mean performance and standard deviation for the TEST set using a 10×10 folds cross-validation process

average results (in the test phase) using **ACC**						
Data	KM_{MV}	PKM_{MV}	PKM_{SNB}	LMT	NBT	SNB
Glass	70.34 ± 8.00	89.32 ± 6.09	95.38 ± 4.66	97.48 ± 2.68	94.63 ± 4.39	$\mathbf{97.75} \pm 3.33$
Pima	65.11 ± 4.17	66.90 ± 4.87	73.72 ± 4.37	$\mathbf{76.85} \pm 4.70$	75.38 ± 4.71	75.41 ± 4.75
Vehicle	37.60 ± 4.10	47.35 ± 5.62	72.21 ± 4.13	$\mathbf{82.52} \pm 3.64$	70.46 ± 5.17	64.26 ± 4.39
Segment	67.50 ± 2.35	80.94 ± 1.93	96.18 ± 1.26	$\mathbf{96.30} \pm 1.15$	95.17 ± 1.29	94.44 ± 1.48
Waveform	50.05 ± 1.05	49.72 ± 3.39	84.04 ± 1.63	$\mathbf{86.94} \pm 1.69$	79.87 ± 2.32	83.14 ± 1.49
Mushroom	89.26 ± 0.97	98.57 ± 3.60	$\mathbf{99.94} \pm 0.09$	98.06 ± 4.13	95.69 ± 6.73	99.38 ± 0.27
PenDigits	73.65 ± 2.09	76.82 ± 1.33	97.35 ± 1.36	$\mathbf{98.50} \pm 0.35$	95.29 ± 0.76	89.92 ± 1.33
Adult	76.07 ± 0.14	77.96 ± 0.41	$\mathbf{86.81} \pm 0.39$	83.22 ± 1.80	79.41 ± 7.34	86.63 ± 0.40
Average	66.19	73.44	88.20	**89.98**	85.73	86.36
average results (in the test phase) using 100 x **AUC**						
Data	KM_{MV}	PKM_{MV}	PKM_{SNB}	LMT	NBT	SNB
Glass	85.72 ± 5.69	96.93 ± 2.84	98.27 ± 2.50	97.94 ± 0.19	98.67 ± 2.05	$\mathbf{99.77} \pm 0.54$
Pima	65.36 ± 5.21	65.81 ± 6.37	78.44 ± 5.35	$\mathbf{83.05} \pm 4.61$	80.33 ± 5.21	80.59 ± 4.78
Vehicle	65.80 ± 3.36	74.77 ± 3.14	91.15 ± 1.75	$\mathbf{95.77} \pm 1.44$	88.07 ± 3.04	87.19 ± 1.97
Segment	91.96 ± 0.75	95.24 ± 0.75	99.51 ± 0.32	$\mathbf{99.65} \pm 0.23$	98.86 ± 0.51	99.52 ± 0.19
Waveform	75.58 ± 0.58	69.21 ± 3.17	96.16 ± 0.58	$\mathbf{97.10} \pm 0.53$	93.47 ± 1.41	95.81 ± 0.57
Mushroom	88.63 ± 1.03	98.47 ± 0.38	$\mathbf{99.99} \pm 0.00$	99.89 ± 0.69	99.08 ± 2.29	99.97 ± 0.02
Pendigits	95.34 ± 0.45	95.84 ± 0.29	99.66 ± 0.11	$\mathbf{99.81} \pm 0.10$	99.22 ± 1.78	99.19 ± 1.14
Adult	73.33 ± 0.65	59.42 ± 3.70	$\mathbf{92.37} \pm 0.34$	77.32 ± 10.93	84.25 ± 5.66	92.32 ± 0.34
Average	80.21	81.96	**94.44**	93.81	92.74	94.29

slightly better results than SNB while both use naive Bayes classifiers locally and PKM_{SNB} is hampered by the fact that $K = J$, the number of classes. Better performance are expected when $K \geq J$. Finally, PKM_{SNB} appears to be slightly superior to SNB, particularly for the datasets which contain highly correlated features, for instance the PenDigits database.

Discussion About Local Models, Complexity and Others Factors: In Sect. 2 in the paragraph about label prediction in predictive clustering, we proposed a list of desirable properties for the local prediction models used in each cluster. We come back to these items denoted from (i) to (vi) in discussing Tables 2 and 3:

i) The performance in prediction are good even for the dataset Glass which contains only 214 instances (90% for training in the 10×10 cross validation (therefore 193 instances)).

ii) The robustness (ratio between the performance in test and training) is given in Table 3 for the Accuracy (ACC) and the AUC. This ratio indicates that there is no significant overfitting. Moreover, by contrast to methods described in [14,15] (about LMT and NBT) our algorithm does not require any cross validation for setting parameters.

iii) The only user parameter is the number of cluster (in this paper $K = J$). This point is crucial to help a non-expert to use the proposed method.

iv) The preprocessing complexity (step 1 of Algorithm 1) is $O(d\,m \log m)$, the k-means has the usual complexity $O(d\,m\,J\,t)$ and the complexity for the creation of the local models is $O(d\,m^* \log m*) + O(K(d\,m^* \log dm^*))$ where d is the number of variables, m the number of instances in the training dataset, m^* is the average number of instances belonging to a cluster. Therefore a fast training time is possible as indicated in Table 3 with time given in seconds (for a PC with Windows 7 enterprise and a CPU: Intel Core I7 6820-HQ 2.70 GHz).

v) Only clusters where the information is sufficient to beat the majority vote contain local model. Table 3 gives the percentage of pure clusters obtained at the end of the convergence the K-Means and the percentage of clusters with a local model (if not pure) when performing the 10×10 cross validation (so over 100 results).

vi) Finally, the interpretation of the PKM$_{SNB}$ model is based on a two-level analysis. The first analysis consists in analyzing the profile of each cluster using histograms. A visualisation of the average profile of the overall population (each bar representing the percentage of instances having a value of the corresponding interval) and the average profile of a given cluster allows to understand why a given instance belongs to a cluster. Then locally to a cluster the variable importance of the local classifier (the weights, W_f, in the SNB classifier) gives a local interpretation.

Table 3. Elements for discussion about local models. (1) Percentage of pure clusters; (2) Percentage of non-pure clusters with a local model; (3) Percentage of non-pure clusters without a local model.

Datasets (#intances)	Robustness		Training Time (s)	Local models			Datasets (#intances)	Robustness		Training Time (s)	Local models		
	ACC	AUC		(1)	(2)	(3)		ACC	AUC		(1)	(2)	(3)
Glass	0.98	0.99	0.07	40.83	23.17	36.0	Waveform	0.97	0.99	0.73	0.00	98.00	2.00
Pima	0.95	0.94	0.05	00.00	88.50	11.5	Mushroom	1.00	1.00	0.53	50.00	50.00	0
Vehicle	0.95	0.97	0.14	07.50	92.25	0.25	Pendigits	0.99	1.00	1.81	0.00	100.00	0
Segment	0.98	1.00	0.85	28.28	66.28	5.44	Adult	1.00	1.00	3.57	0.00	100.00	0

The results of our experiments and the elements (i) to (vi) show that the algorithm PKM$_{SNB}$ is interesting with regards to several aspects. (1) Its predictive performance are comparable to those of the best competing supervised classification methods, (2) it doesn't require cross validation, (3) it deals with the missing values, (4) it operates a features selection both in the clustering step and during the building of the local models. Finally, (5) it groups the categorical features into modalities, thus allowing one to avoid using a complete disjunctive coding which involves the creation of large vectors. Otherwise this disjunctive coding could complicate the interpretation of the obtained model.

The reader may find a supplementary material here: https://bit.ly/2T4VhQw or here: https://bit.ly/3a7xmFF. It gives a detailed example about the interpretation of the results and some comparisons to others predictive clustering algorithms as COBRA or MPCKmeans.

4 Conclusion and Perspectives

We have shown how to modify a distance-based clustering technique, such as k-means, into a predictive clustering algorithm. Moreover the learned representation could be used by other clustering algorithms. The resulting algorithm PKM_{SNB} exhibits strong predictive performances most of the time as the state of the art but with the benefit of not having any parameters to adjust and therefore no cross validation to compute. The suggested algorithm is also a good support for interpretation of the data. Better performances can still be expected when the number of clusters is higher than the number of classes. One goal of a work in progress it to find a method that would automatically discover the optimal number of clusters. In addition, we are developing a tool to help visualize the results allowing the navigation between clusters in order to view easily the average profiles and the importance of the variables locally for each cluster.

References

1. Al-Harbi, S.H., Rayward-Smith, V.J.: Adapting k-means for supervised clustering. Appl. Intell. **24**(3), 219–226 (2006)
2. Arthur, D., Vassilvitskii, S.: K-means++: the advantages of careful seeding. In: Proceedings of the Eighteenth Annual ACM-SIAM Symposium on Discrete Algorithms, pp. 1027–1035 (2007)
3. Kim, B., Varshney, K.R., Weller, A.: Workshop on human interpretability in machine learning (WHI 2018). In: Proceedings of the 2018 ICML Workshop (2018)
4. Bilenko, M., Basu, S., Mooney, R.J.: Integrating constraints and metric learning in semi-supervised clustering. In: Proceedings of the Twenty-First International Conference on Machine Learning (ICML) (2004)
5. Blockeel, H., Dzeroski, S., Struyf, J., Zenko, B.: Predictive Clustering. Springer, New York (2019)
6. Bouchard, G., Triggs, B.: The tradeoff between generative and discriminative classifiers. In: IASC International Symposium on Computational Statistics (COMPSTAT), pp. 721–728 (2004)
7. Boullé, M.: A Bayes optimal approach for partitioning the values of categorical attributes. J. Mach. Learn. Res. **6**, 1431–1452 (2005)
8. Boullé, M.: MODL: a Bayes optimal discretization method for continuous attributes. Mach. Learn. **65**(1), 131–165 (2006)
9. Boullé, M.: Compression-based averaging of selective Naive Bayes classifiers. J. Mach. Learn. Res. **8**, 1659–1685 (2007)
10. Cevikalp, H., Larlus, D., Jurie, F.: A supervised clustering algorithm for the initialization of RBF neural network classifiers. In: Signal Processing and Communication Applications Conference, June 2007. http://lear.inrialpes.fr/pubs/2007/CLJ07
11. Eick, C.F., Zeidat, N., Zhao, Z.: Supervised clustering - algorithms and benefits. In: International Conference on Tools with Artificial Intelligence, pp. 774–776 (2004)

12. Flach, P.: Machine Learning: The Art and Science of Algorithms That Make Sense of Data. Cambridge University Press, Cambridge (2012)
13. Hand, D.J., Yu, K.: Idiot's Bayes-not so stupid after all? Int. Stat. Rev. **69**(3), 385–398 (2001)
14. Kohavi, R.: Scaling up the accuracy of Naive-Bayes classifiers: a decision-tree hybrid. In: International Conference on Data Mining, pp. 202–207. AAAI Press (1996)
15. Landwehr, N., Hall, M., Frank, E.: Logistic model trees. Mach. Learn. 59(1–2) (2005)
16. Langley, P., Sage, S.: Induction of selective Bayesian classifiers. In: Proceedings of the Tenth International Conference on Uncertainty in Artificial Intelligence, pp. 399–406. Morgan Kaufmann Publishers Inc., San Francisco (1994)
17. Lemaire, V., Alaoui Ismaili, O., Cornuéjols, A.: An initialization scheme for supervized k-means. In: International Joint Conference on Neural Networks (2015)
18. Lichman, M.: UCI machine learning repository (2013)
19. Manning, C.D., Raghavan, P., Schütze, H.: Introduction to Information Retrieval. Cambridge University Press, New York (2008)
20. Meilă, M., Heckerman, D.: An experimental comparison of several clustering and initialization methods. In: Conference on Uncertainty in Artificial Intelligence, pp. 386–395. Morgan Kaufmann Publishers Inc. (1998)
21. Milligan, G.W., Cooper, M.C.: A study of standardization of variables in cluster analysis. J. Classif. 5(2), 181–204 (1988)
22. Salperwyck, C., Lemaire, V.: Learning with few examples: an empirical study on leading classifiers. In: International Joint Conference on Neural Networks (2011)
23. Van Craenendonck, T., Dumancic, S., Van Wolputte, E., Blockeel, H.: COBRAS: fast, iterative, active clustering with pairwise constraints. In: Proceedings of Intelligent Data Analysis (2018)

Automatic Detection of Sexist Statements Commonly Used at the Workplace

Dylan Grosz[1,2]([✉]) and Patricia Conde-Cespedes[2]

[1] Stanford University, Stanford, CA 94305, USA
dgrosz@stanford.edu
[2] ISEP (Institut supérieur d'électronique de Paris), Paris, France
patricia.conde-cespedes@isep.fr

Abstract. Detecting hate speech in the workplace is a unique classification task, as the underlying social context implies a subtler version
of conventional hate speech. Applications regarding a state-of-the-art
workplace sexism detection model include aids for Human Resources
departments, AI chatbots and sentiment analysis. Most existing hate
speech detection methods, although robust and accurate, focus on hate
speech found on social media, specifically Twitter. The context of social
media is much more anonymous than the workplace, therefore it tends to
lend itself to more aggressive and "hostile" versions of sexism. Therefore,
datasets with large amounts of "hostile" sexism have a slightly easier
detection task since "hostile" sexist statements can hinge on a couple
words that, regardless of context, tip the model off that a statement is
sexist. In this paper we present a dataset of sexist statements that are
more likely to be said in the workplace as well as a deep learning model
that can achieve state-of-the art results. Previous research has created
state-of-the-art models to distinguish "hostile" and "benevolent" sexism
based simply on aggregated Twitter data. Our deep learning methods,
initialized with GloVe or random word embeddings, use LSTMs with
attention mechanisms to outperform those models on a more diverse,
filtered dataset that is more targeted towards workplace sexism, leading
to an F1 score of 0.88.

Keywords: Hate speech · Natural Language Processing · Sexism ·
Workplace · LSTM · Attention mechanism

1 Introduction

Not long ago, women in the U.S. were not entitled to vote, yet in 2016 the first
woman in history was nominated to compete against a male opponent to become
President of the United States. A similar situation took place in France in 2017
when Marine Le Pen faced Emmanuel Macron in the runoff election. Is the gap
between male and female opportunities in the workplace changing? A recent

---

Supported by Stanford University and ISEP.

W. Lu and K. Q. Zhu (Eds.): PAKDD 2020 Workshops, LNAI 12237, pp. 104–115, 2020.
https://doi.org/10.1007/978-3-030-60470-7_11

study conducted by McKinsey and Company (2017) [14], reveals the gaps and patterns that exist today between women and men in corporate America. The results of the study reveal that many companies have not made enough positive changes, and as a result, women are still less likely to get a promotion or get hired for a senior level position. Some key findings from this study include, for instance:

- Corporate America awards promotions to males are about 30% higher rate than women in the early stages of their careers.
- Women compete for promotions as often as men, yet they receive more resistance.

Mary Brinton [5], sociology professor at Harvard University and instructor of Inequality and Society in Contemporary Japan, points out that although men and women are now on an equal playing field in regard to higher education, inequality persists. Furthermore, some women who occupy important positions or get important achievements suffer from sexism at their workplace. One can mention, for example, the incident that took place in December 2018 during the *Ballon d'Or* ceremony when host Martin Solveig asked the young Norwegian football player Ada Hegerberg, who was awarded the inaugural women's Ballon d'Or, was asked: *"Do you know how to twerk?"* [1]. Even more recently, a young scientist Katie Bouman, a postdoctoral fellow at Harvard, was publicly attributed to have constructed the first algorithm that could visualize a black hole [18]. Unfortunately, this event triggered a lot of sexist remarks on social media questioning Bouman's role in the monumental discovery. For instance, a YouTube video titled *Woman Does 6% of the Work but Gets 100% of the Credit* garnered well over 100K views. Deborah Vagins, member of the American Association of University Women, emphasized that women continue to suffer discrimination, especially when a woman works in a male-dominated field (the interested reader can see [7,11,17,24]). Another relevant example is physicist Alessandro Strumia University of Pisa who was suspended from CERN (Conseil européen pour la recherche nucléaire) for making sexist comments during a presentation claiming that physics was becoming sexist against men. *"the data doesn't lie-women don't like physics"*, *"physics was invented and built by men"* were some of the expressions he used [19,20].

All these examples bring out that the prejudicial and discriminatory nature of sexist behavior unfortunately pervades nearly every social context, especially for women. This phenomenon leads sexism to manifest itself in social situations whose stakes can lay between the anonymity of social media (twitter, Facebook, youtube) and the relatively greater social accountability of the workplace.

In this paper, based on recent Natural Language Processing (NLP) and deep learning techniques, we built a classifier to automatically detect whether or not statements commonly said at work are sexist. We also manually built a dataset of sentences containing neutral and sexist content.

Section 2 presents a literature review of automatic hate speech detection methods using NLP methods. Moving on to our novel work and contributions, Sect. 3 describes the unique dataset used for our experimental results, one which we hope future research will incorporate and improve upon. Next, Sect. 4 describes the methods used for building our classifier. Then, in Sect. 5 we present the experimental results. Finally, Sect. 6 presents our conclusion and perspectives of this study.

The code and dataset that are discussed in this paper will be available at github.com/dylangrosz.

2   Related Works

For many years, fields such as social psychology have deeply studied the nature and effects of sexist content. The many contexts where one can find sexism are further nuanced by the different forms sexist speech can take. In 1996 Glick and Fiske [9] devised a theory introducing the concept of *Ambivalent Sexism*, which distinguishes between a *benevolent* and *hostile* sexism. Both forms involve issues of paternalism, predetermined ideas of women's societal roles and sexuality; however, *benevolent* sexism is superficially more positive and benign in nature than *hostile* sexism, yet it can carry similar underlying assumptions and stereotypes. The distinction between the two types of "sexisms" was extended recently by Jha and Mamidi (2017) [12]. The authors characterized hostile sexism by an explicitly negative attitude whereas they remarked benevolent sexism is more subtle, which is why their study was focused on identifying this less pronounced form of sexism.

Jha and Mamidi, 2017 [12] have successfully proposed a method that can disambiguate between these benevolent and hostile sexisms, suggesting that there are perhaps detectable traits of each category. Through training SVM and Sequence-to-Sequence models on a database of hostile, benevolent and neutral tweets, the two models performed with an F1 score of 0.80 for detecting benevolent sexism and 0.61 for hostile sexism. These outcomes are quite decent considering that the little preprocessing left a relatively unstructured dataset from which to learn. With regards to the context presented in our research, the workplace features much more formal and subversive sexism as compared to that found on social media, so such success in detecting benevolent sexism is useful for our purpose.

Previous research has also found some success on creating models that can disambiguate various types of hate speech and offensive language in the social media context. A corpus of sexist and racist tweets was debuted by Waseem and Hovy (2016) [25]. This dataset was further labeled as *Hostile* and *Benevolent* versions of sexism by Jha and Mamidi (2017) [12] which Badjatiya et al. (2017) [2], Pitsilis et al. (2018) [22] and Founta et al. (2018) [8] all use as a central training dataset in their research, each attempting to improve classification results with various model types. Waseem and Hovy (2016) [25] experimented with simpler learning techniques such as logistic regression, yielding an F1 score of 0.78

when classifying offensive tweets. Later studies by [2] experimented with wide varieties of deep learning architectures, but success seemed to coalesce around ensembles of Recurrent Neural Network (RNN), specifically Long Short-Term Memory (LSTM) classifiers. Results for these studies featured F1 scores ranging from 0.90 to 0.93 after adding in various boosting and embedding techniques.

For this research, several models were employed to figure out which best predicted workplace sexism given the data. While the more basic models relied on some form of logistic regression, most other tested models employed deep learning architectures. Of these deep learning models, the simplest used a unidirectional LSTM layers, while the most complex employed a bidirectional LSTM layer with a single attention layer [3], allowing the model to automatically focus on relevant parts of the sentence. Most of these models used GloVe embedding, a project meant to place words in a vector space with respect to their co-occurrences in a corpus [21]. Some models experimented with Random Embedding, which just initializes word vectors to random values so as to not give the deep learning model any given "intuition" before training.

Among all this related research, none specifically considered the specific context of the workplace. Rather, most of them share a curated dataset of 16K tweets from Twitter in their hate speech detection and classification tasks. Given the substantial difference in datasets and contexts, our paper proposes a new dataset of sexist statements in the workplace and an improved companion deep learning method that can achieve results akin to these previous hate speech detection tasks.

3   Dataset Description

The dataset used in model training and testing features more than 1100 examples of statements of workplace sexism, roughly balanced between examples of certain sexism and ambiguous or neutral cases (labeled with a "1" and "0" respectively). Though this dataset features some sexist statements from Twitter, it differs from previous Twitter datasets in hate speech detection research. Previous Twitter datasets were collected via keywords and hashtags, which does not port well over to workplace speech since the nature of the dataset suffers greatly from:

1. Over-representing rare sexist scenarios (e.g., the name Kat is regarded as sexist since she was a figure many people directed sexist comments during Season 6 of My Kitchen Rules (#MKR)).
2. Unnatural amplification of certain phraseology through retweeting since all collected retweets just reproduce the original tweet attached with the username of the user who retweeted.
3. Learning Twitter-specific tokens, especially internet slang and hashtags, which should be left unlearned with respect to the workplace context.

The Twitter portion of our dataset alleviates the first issue by filtering out these rare scenarios through generalizing certain tweets (e.g., many usages of "Kat" are converted to "she" or "her"). The second issue is resolved through removing duplicates of tweet bodies and preserving only the original tweet. The final issue was resolved manually by writing out or removing hashtags (the latter occurs if it happens at the end of the tweet and has no additional contextual relevance) and converting casual slang to its more formal, work-appropriate version (e.g., "u" becomes "you"). While 55% of the dataset includes these generic tweets of "benevolent" sexism, other sources of workplace-related sexist speech are included to keep the source contexts of the workplace statements diversified in order to reduce overfitting on confounding keywords and phrase constructions:

- 55% - A manually filtered subset of a Twitter hate speech dataset created by [25]
- 25% - A manually filtered subset of work-related quotes [10]
- 20% - Miscellaneous press quotes and faculty/student submissions [6, 16, 23, 26]

NOTE: Manual data selection and filtering was done by Grosz (male) and spot checked by Conde-Cespedes (female).

Examples of certain workplace sexism must be both conceivable in a workplace environment and somewhat professional in nature. The latter requirement is a bit loose since workplace sexism can include obvious and/or "hostile" sexism. Such examples include:

- ''Women always get more upset than men.''
- ''The people at work are childish. it's run by women and when women dont agree to something, oh man.''
- ''I'm going to miss her resting bitch face.''
- ''Seeing as you two think this is a modelling competition, I give you two a score of negative ten for your looks.''

Examples of ambiguous or neutral cases include:

- ''No mountain is high enough for a girl to climb.''
- ''The Belgian bar near the end of the road was a great spot to go after work''
- ''It seems the world is not ready for one of the most powerful and influential countries to have a woman leader. So sad.''
- ''Can you explain why what she described there is wrong?''

Some ethical concerns can arise in implicitly defining sexism via these datapoints. Since sexism is mostly directed towards women in the collected data, subsequent modelling will reflect that imbalance through having a more nuanced understanding and a higher confidence in labelling new examples of women-directed sexism than man-directed sexism. As a neutral counterweight to the bias, a good proportion of positive and negative examples are generic enough to detect a woman be sexist towards a man. For example, the model detects a "he" and a "she" in the statement "He thinks she should consult her gender before working here." An ideal model would give less weight to the order of the subjects, but should be able to deduce that if the predicate of the statement is somewhat negative and is paired with a he vs. she set-up, the model will lean more towards predicting the phrase as being sexist regardless of to which gender the statement was levied.

Of the more than 1100 total statements, 55% are labeled as sexist ("1") while 45% are labeled as ambiguous/neutral ("0"). The dataset is publicly available at github.com/dylangrosz.

4   Description of the Classification Method

We experimented with various classification methods to see which would yield the best results. Our models take some inspiration from previous state-of-the-art hate speech classification models. We considered four groups of model versions, denoted from V1 to V4. All of these models take in word embeddings for each word in a sentence, initialized randomly, through GloVe or through GN-GloVe (a gender debiased version of GloVe) [27]. After propogating through the model, outputs a binary classification pertaining to its status as sexist.

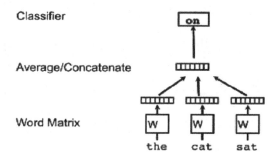

Fig. 1. General architecture of models V1: a simple classifier based on word embeddings. (the image was taken from [13])

In each group, there are sub-versions that experiment with different sub-architectures. In total, this research considers seven model versions. In Table 1 we present a summary of the performance metrics for each model in terms of *recall, precision* and *F1 score*.

- **Version 1** (V1) of the models (seen in Fig. 1) are a class of models using non-deep learning techniques with learned embeddings, which can serve as a baseline to which deep learning models can compare. Model V1a uses GloVe word embeddings to calculate an average embedding of the statement, while Model V1b uses GloVe embeddings, but instead of calculating the average and training a logistic regression classifier like in V1a, it trains a Gradient

Boosted Decision Tree classifier. These models established a baseline F1 level of around 0.83.

- **Version 2** (V2) of the models employs a LSTM deep learning architecture (seen in Fig. 2). After an embedding layer initialized on GloVe, inputs are propagated through two unidirectional LSTM layers. In theory, this model should be able to perceive more nuanced phrases in context. For example, the V1 model would perceive a phrase such as "not pretty" individually; the LSTM construction allows the model to be able to perceive this "not pretty" as the opposite of the "pretty" in the context of its classification task. This construction had similar results to V1, also yielding a F1 of 0.83.

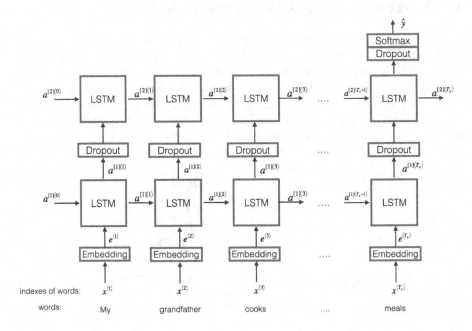

Fig. 2. General architecture of models V2: A LSTM classifier based on GloVe word embeddings

- **Version 3** (V3) of the models (seen in Fig. 3) is very similar to V2, but it substitutes the unidirectional LSTM layers with bidirectional LSTM layers. A random embedding scheme was tested in Version 3a, GloVe for Version 3b and GN-GloVe for Version 4c. This change should allow the model to read the phrases both forwards and backwards to better learn their nuanced meanings. A phrase such as "women and men are work great together" might be more likely to be labeled as sexist by V2 due to the presence of "women and men" (which appears in many other obviously) and its ensuing influence on classification. With a separate portion of the LSTM layer devoted to "reading" the statement in the other direction, it will read "work great together" first, which will influence the classification to be non-sexist. On balance, this architecture might better perceive the nuance of certain sexist or non-sexist statements. The introduction of bidirectional layers yielded a slightly improved F1 of 0.85.

– **Version 4** (V4) of the models (seen in Fig. 4) employs the same architecture as V3. However, it adds a simple attention mechanism over the embedding input layer in order to focus on the significance of individual words out of context. Like in V3, random embedding was tested in Version 4a, GloVe for Version 4b and GN-GloVe for Version 4c. For example, the model tends to over-label statements including "women and men" as sexist, since it implies a comparison which usually invokes sexist stereotypes; however, there are many cases where "women and men" are followed by an undeniably neutral clause, as seen in the example statement "men and women should like this product." The attention mechanism seeks to learn that "should like this product," usually regardless of context, means a workplace statement is not sexist. As a result of this greater understanding of statements' nuance, this model fared best with a F1 of 0.88.

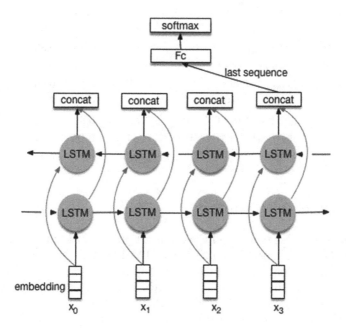

Fig. 3. Architecture of the BiLSTM component of model V3: A deep learning model with a bidirectional LSTM layer that can understand sentences both forwards and backwards [15]

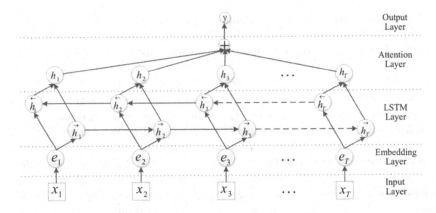

Fig. 4. General architecture of the BiLSTM+Attention component of model V4: A bidirectional LSTM classifier that propagates into an attention layer [28]

Table 1. Model performances

Model	Description	Precision	Recall	F1 score
V1a	GloVe+Logistic Regression	0.81	0.86	0.83
V1b	GloVe+GBDT	0.82	0.84	0.83
V2	GloVe+LSTM	0.80	0.87	0.83
V3a	Random+BiLSTM	0.80	0.77	0.79
V3b	GloVe+BiLSTM	0.82	0.89	0.85
V3c	GN-GloVe+BiLSTM	0.82	0.89	0.85
V4a	Random+BiLSTM+Attn	0.71	0.81	0.76
V4b	GloVe+BiLSTM+Attn	**0.84**	**0.93**	**0.88**
V4c	GN-GloVe+BiLSTM+Attn	0.82	0.92	0.87

5   Experimental Results

Though the simple V1 model and the unidirectional LSTM initialized on GloVe posted similar F1 scores, changing the LSTM layers to be bidirectional and adding a simple attention mechanism substantially improved the F1 score to 0.88. Though promising in previous research, initializing with random embeddings led to poor F1 scores and irreconcilable overfitting.

While the pretrained GloVe embedding led to the best results, a common criticism of such pretrained embeddings is the possibility that they can assume certain human biases, such as gender bias. However, training the model on a gender neutral version of GloVe (GN-GloVe) showed no significant improvement to performance [27], possibly due to either a slight advantage on having mathematically embedded gender biases or the irrelevance of analogical gender biases with respect to this task. However, gender neutral word embeddings may prove promising as underlying detection tasks evolve and more research comes out regarding the debiasing of generic, complex word embeddings like GloVe, as opposed to targeted, simpler word embeddings like Google News' word2vec [4].

Even for the best model, persistent issues include an over-aggressive labeling of sentences that include the phrase "women and men," slight overfitting despite Dropout layers and recall slightly outperforming precision (the models over-labeled statements as sexist as a whole).

For optimal training and testing, V2, the V3s, and the V4s featured layers with sizes between 64 and 128. There are also Dropout layers between each LSTM layer to reduce overfitting. The model was then compiled to optimize via binary cross entropy and an 'adam' optimizer.

6   Conclusion and Future Works

The GloVe+BiLSTM+Attn model's F1 score of 0.88 shows that with the slightly different deep learning methods shown in this paper, a F1 score that is at the level of previous sexist detection research is attainable. This performance must also be taken into context with this task's added limitation of constraining all data to be in a workplace context; this type of data leans much more into the category of the more nuanced, subtler "benevolent" sexism.

With a larger dataset, the GloVe+BiLSTM+Attn will be more able to abstract from the data and learn the most generalized and accurate model possible. Although the dataset size is the most obvious culprit for being the bottleneck for further F1 improvement, there are also more possible, complex and novel deep learning architectures that can be explored and tested on the dataset, including boosting techniques, other pretrained word embedding and more sophisticated attention mechanisms. This final improvement could pose an especially ripe area for including more explainablity and understanding since attention mechanisms can allow one to peer into the key words and phrases the model focuses on when tagging statements as sexist or not.

The dataset presented in this paper could also be used as a basis for unsupervised learning tasks via clustering to reverse-engineer more nuanced types of sexism, as there can possibly be subclasses of both hostile and benevolent sexism that upon discovery could help sociological reframings of this problem as well as helping understand this task itself.

The dataset used in this research, though large enough to produce substantially robust models in workplace sexism detection, must always grow in order to capture most keywords and phrase structures found in workplace sexism. Despite the challenges posed by the current state of the dataset and model, state-of-the-art results were attained. As this dataset of sexist workplace statements grows through a crowdsourced effort, the performance of this model will improve as well.

References

1. Aarons, E.: Ada Hegerberg: First Women's Ballon d'Or Marred as Winner is Asked to Twerk. The Guardian (2018). https://www.theguardian.com/football/2018/dec/03/ballon-dor-ada-hegerberg-twerk-luka-modric
2. Badjatiya, P., Gupta, S., Gupta, M., Varma, V.: Deep learning for hate speech detection in tweets. In: 26th International Conference on World Wide Web Companion, pp. 759–760 (2017)
3. Bahdanau, D., Cho, K., Bengio, Y.: Neural machine translation by jointly learning to align and translate. In: International Conference on Learning Representations (2015)
4. Bolukbasi, T., et al.: Man is to computer programmer as woman is to homemaker? Debiasing word embeddings. In: NIPS (2016)
5. Brinton, M.: Gender inequality and women in the workplace. Harvard Summer School Review, May 2017. https://www.summer.harvard.edu/inside-summer/gender-inequality-women-workplace
6. Chira, S., Milord, B.: "'Is there a man i can talk to?': stories of sexism in the workplace". The New York times. The New York Times, June 2017. www.nytimes.com/2017/06/20/business/women-react-to-sexism-in-the-workplace.html
7. Elfrink, T.: Trolls hijacked a scientist's image to attack Katie Bouman. They picked the wrong astrophysicist. The Washington Post, p. 2019, April 2019. https://www.washingtonpost.com/nation/2019/04/12/trolls-hijacked-scientists-image-attack-katie-bouman-they-picked-wrong-astrophysicist
8. Founta, A.M., Chatzakou, D., Kourtellis, N., Blackburn, J., Vakali, A., Leontiadis, I.: A unified deep learning architecture for abuse detection. CoRR (2018). http://arxiv.org/abs/1802.00385
9. Glick, P., Fiske, S.T.: The ambivalent sexism inventory: differentiating hostile and benevolent sexism. J. Pers. Soc. Psychol. **3**, 491–512 (1996)
10. Goel, S., Madhok, R., Garg, S.: Proposing contextually relevant quotes for images. In: Pasi, G., Piwowarski, B., Azzopardi, L., Hanbury, A. (eds.) ECIR 2018. LNCS, vol. 10772, pp. 591–597. Springer, Cham (2018). https://doi.org/10.1007/978-3-319-76941-7_49
11. Griggs, M.B.: Online trolls are harassing a scientist who helped take the first picture of a black hole. The Verge, April 2019. https://www.theverge.com/2019/4/13/18308652/katie-bouman-black-hole-science-internet
12. Jha, A., Mamidi, R.: When does a compliment become sexist? Analysis and classification of ambivalent sexism using Twitter data. In: Proceedings of the Second Workshop on NLP and Computational Social Science edn. Association for Computational Linguistics (2017)
13. Kabbani, R.: Checking Eligibility of Google and Microsoft Machine Learning Tools for use by JACK e-Learning System (2016)
14. Krivkovich, A., Robinson, K., Starikova, I., Valentino, R., Yee, L.: Women in the Workplace. McKinsey & Company (2017). https://www.mckinsey.com/featured-insights/gender-equality/women-in-the-workplace-2017
15. Lee, C.: Understanding bidirectional RNN in pytorch. Towards Data Science, November 2017. http://towardsdatascience.com/understanding-bidirectional-rnn-in-pytorch-5bd25a5dd66
16. McCormack, C.: 18 sexist phrases we should stop using immediately. MSN (2018). https://www.msn.com/en-us/Lifestyle/smart-living/18-sexist-phrases-we-should-stop-using-immediately/ss-BBLgg1E/

17. Mervosh, S.: How Katie Bouman accidentally became the face of the black hole project. The New York Times, April 2019. https://www.nytimes.com/2019/04/11/science/katie-bouman-black-hole.html

18. @MIT_CSAIL, April 2019. https://twitter.com/MIT_CSAIL/status/1116020858282180609

19. BBC News: Cern scientist alessandro strumia suspended after comments. BBC News, October 2018. https://www.bbc.com/news/world-europe-45709205

20. Palus, S.: We annotated that horrible article about how women don't like physics. Slate, March 2019. https://slate.com/technology/2019/03/women-dont-like-physics-article-annotated.html

21. Pennington, J., Socher, R., Manning, C.D.: Glove: global vectors for word representation (2014)

22. Pitsilis, G., Ramampiaro, H., Langseth, H.: Detecting offensive language in tweets using deep learning. CoRR (2018)

23. Priestley, A.: Six common manifestations of everyday sexism at work. Smart-Company, October 2017. www.smartcompany.com.au/people-human-resources/six-common-manifestations-everyday-sexism-work/

24. Resnick, B.: Male scientists are often cast as lone geniuses. Here's what happened when a woman was. Vox, April 2019. https://www.vox.com/science-and-health/2019/4/16/18311194/black-hole-katie-bouman-trolls

25. Waseem, Z., Hovy, D.: Hateful symbols or hateful people? Predictive features for hate speech detection on Twitter. In: Proceedings of the NAACL Student Research Workshop at Association for Computational Linguistics 2016, pp. 88–93 (2016)

26. Wolfe, L.: Sexist comments and quotes made by the media. The Balance Careers, March 2019. www.thebalancecareers.com/sexist-comments-made-by-media-3515717

27. Zhao, J., Zhou, Y., Li, Z., Wang, W., Chang, K.W.: Learning gender-neutral word embeddings. In: ACL, pp. 4847–4853 (2018)

28. Zhou, P., et al.: Attention-based bidirectional long short-term memory networks for relation classification. In: ACL (2016). http://www.aclweb.org/anthology/P16-2034

Hybrid Dimensionality Reduction Technique for Hyperspectral Images Using Random Projection and Manifold Learning

Alkha Mohan[(✉)] and M. Venkatesan

Department of Computer Science and Engineering,
National Institute of Technology Karnataka,
Srinivasnagar PO, Surathkal, Mangalore 575025, India
mohan.alkha@gmail.com, venkisakthi77@gmail.com

Abstract. Hyperspectral images (HSI) are contiguous band images having hundreds of bands. However, most of the bands are redundant and irrelevant. Curse of dimensionality is a significant problem in hyperspectral image analysis. The band extraction technique is one of the dimensionality reduction (DR) method applicable in HSI. Linear dimensionality reduction techniques fail for hyperspectral images due to its nonlinearity nature. Nonlinear reduction techniques are computationally complex. Therefore this paper introduces a hybrid dimensionality reduction technique for band extraction in hyperspectral images. It is a combination of linear random projection (RP) and nonlinear technique. The random projection method reduces the dimensionality of hyperspectral images linearly using either Gaussian or Sparse distribution matrix. Sparse random projection (SRP) is computationally less complex. This reduced image is fed into a nonlinear technique and performs band extraction in minimal computational time and maximum classification accuracy. For experimental analysis of the proposed method, the hybrid technique is compared with Kernel PCA (KPCA) using different random matrix and found a promising improvement in results for their hybrid models in minimum computation time than classic nonlinear technique.

Keywords: Hyperspectral images · Curse of dimensionality · Band reduction · Random projection · Manifold learning

1 Introduction

Remote sensing image analysis became an emerging area nowadays. Various types of remote sensing images such as RADAR images, multispectral images, hyperspectral images are using for different applications based on their spatial resolution, spectral resolution, data collection techniques, etc. The reflected energy from any material has a unique footprint, which is called the spectral

W. Lu and K. Q. Zhu (Eds.): PAKDD 2020 Workshops, LNAI 12237, pp. 116–127, 2020.
https://doi.org/10.1007/978-3-030-60470-7_12

signature of that material. This spectral signature [6] is used to discriminate various objects on the earth's surface. Multispectral images are having around 3–10 number of spectral components which captured beyond the visible spectrum. As the number of bands in multispectral images increases, the bandwidth decreases, and reconstruction of the complete spectral signature became easy. Thus researchers started the use of hyperspectral images for crop monitoring [14], change detection [2,13], weather prediction, etc. due to its spectral-spatial features and a large number of bands.

Hyperspectral images are contiguous band images that look like a 3D image cube. Two dimensions in this cube are spatial coordinates, and the third dimension is spectral bandwidth. Thus a single hyperspectral image is the collection of hundreds of images captured in different spectral bandwidth. The contiguous spectral property of these images helps to construct the spectral signature and discriminate objects efficiently. Significant challenges in hyperspectral image classification are lack of ground truth images, limited amount spectral library. Even though a large number of bands improves the material discrimination, it leads to the curse of dimensionality due to the less number of labeled samples and redundancy in bands [15]. Therefore dimensionality reduction in hyperspectral images is essential to improve the classification accuracy and reduce the computational complexity.

Dimensionality reduction mainly classified into three types, namely supervised, semi-supervised, and unsupervised. Supervised DR is not valid for hyperspectral images due to less availability of hyperspectral ground truth. High dimensional natural images posses fewer degrees of freedom. For hyperspectral images, the degree is determined by the number of bands similar to features in other datasets. Therefore we have two methods of DR for hyperspectral images named band selection and band extraction similar to feature selection and feature extraction.

The spectral difference between nearby bands in hyperspectral images is very less. Therefore some bands are exactly similar to its neighboring bands. Band selection techniques find redundancy in bands and select unique bands from the input image [4]. These bands have the same property of input image, whereas band extraction converts the input image into a lower-dimensional space to avoid higher dimensionality issues. Band extraction methods are more renowned than band selection techniques.

Band extraction methods are of two types, namely linear and nonlinear DR (manifold learning) techniques [4]. Linear dimensionality reduction techniques produce a linear low dimensional mapping of high dimensional input data. Commonly used linear band extraction methods are principal component analysis (PCA) [10,11], linear discriminant analysis (LDA) [10], minimum noise fraction (MNF) [9], etc. PCA tries to maximize the data variance, whereas the goal of LDA is to maximize the separation between classes of data. Most of the linear DR techniques preserve up to second-order statistics of input data and not considering the higher-order statistics.

Natural images are nonlinear, and linear reduction techniques are fails in their processing. They are not preserving nonlinearity in input images. Later, researchers started using nonlinear band extraction techniques such as KPCA [1,5], nonlinear independent component analysis (ICA) [17], LLE [1,12], Isomap [1], etc. in hyperspectral images. KPCA is an extension of linear PCA, which applies band extraction on feature space instead of the input image and feature space depend on the selection of a kernel. LLE and Isomap are almost similar methods, and they focus on the internal distribution of data. These methods serve the nonlinearity of natural images and preserve the higher-order statistics. However, the major problems of these nonlinear techniques are their computational complexity. In KPCA, the size of the kernel matrix is $n \times n$, n is the total number of pixels in the input image. Based on the selection of different kernel functions, the computation of the kernel coefficient became more complex. For example, if we select the Gaussian kernel function, we have to calculate the squared Euclidean distance between each pair of pixels in the image.

Similarly, the nearest neighbor finding in LLE and Isomap also calculate the distance between all pair of pixels in the input image. SSNG is another DR technique used in the hyperspectral image by considering both spectral and spatial pixels. This method also creates a neighborhood graph using distance measure [8]. This distance calculation leads to the computational complexity of $O(Dn^2)$, where D is the number of bands in the input image.

The proposed methodology aims to reduce the computational complexity of nonlinear dimensionality reduction techniques in hyperspectral images. This article proposed a hybrid dimensionality reduction technique, which is the combination of a linear and nonlinear DR. Linear reduction technique reduces the input dimension D to k, and this band reduced image take as input for nonlinear technique. Thus this hybrid reduction technique reduces the computational complexity. The proposed method uses random projection as a linear dimensionality reduction technique. Since the random projection matrix is sparse, its computation is less complicated compared with other techniques.

The remaining portion of the paper organized as follows: Sect. 2 gives a brief description of all methodology used in this work and explains the proposed method with the algorithm. Section 3 includes dataset description, experimental setup, and performance evaluation measures. The result analysis of the proposed technique and findings are included in Sect. 4, and the paper is concluded in Sect. 5.

2   Hybrid Band Extraction Methodology

A hyperspectral image is represented as an image cube of size $m \times n \times D$, where m and n are the height and width of each image (representing spatial coordinate) and D is the number of bands (representing spectral dimension). In dataset representation, each data $X = \{x_1, x_2, \ldots, x_N\} \in \mathbb{R}^{D \times N}$ belongs to some class c and N is the total number of pixels ($N = m * n$).

The proposed hybrid band extraction is a combination of linear Random Projection method and manifold learning techniques. Since nonlinear manifold techniques are computationally complex, we apply a linear band extraction technique that map input image of dimension D into k and then the nonlinear technique is applied on k dimension data. This section focus on the detailed description of each band extraction method with the proposed algorithm.

2.1 Random Projection

Random projection [7,16] is a dimensionality reduction technique, which maps the high dimensional data into a lower dimension by preserving the distance between data points. Suppose the input data X having size $n \times d$ random projection map X into a lower-dimensional data Y of size $n \times k$ using a random projection matrix $R^{d \times k}$. The computational complexity of random projection is $O(knd)$.

$$Y^{n \times k} = X^{n \times d} R^{d \times k} \tag{1}$$

The idea behind random projection is Johnson Lindenstrauss lemma [7]: Suppose we have an arbitrary matrix $X \in \mathbb{R}^{n \times d}$. Given any $\epsilon > 0$, there is a mapping $f : R^d \rightarrow R^k$, for any $k \geq O\frac{logn}{\epsilon^2}$, such that, for any two rows $x_i, x_j \in X$, we have

$$(1 - \epsilon)||x_i - x_j||^2 \leq ||f(x_i) - f(x_j)||^2 \leq (1 + \epsilon)||x_i - x_j||^2 \tag{2}$$

Equation (2) states that while converting input data from d dimension to k dimension, its Euclidean distance is preserved with a factor of $1 \pm \epsilon$.

Gaussian random projection (GRP) is one of the most commonly used random projections. Here the random matrix R generated from Gaussian distribution, which satisfies Orthogonality and normality. For a $m \times n$ random matrix, the time complexity for random projection to k dimension is $O(mnk)$. Time complexity reduces to $O(snk)$ if the random matrix has only s number of non-zero elements. Thus Sparse random projection (SRP) evolved with a new random matrix R having more zero elements [3,7].

2.2 Kernel PCA

Kernel PCA is an extension of PCA in nonlinear space, capable of capturing the higher-order statistics of data. Input is transformed into a new feature space, where the discrimination power of each instant from other is very high. Convert data in $\mathbb{R}^D$ to feature space F by applying a nonlinear transformation ϕ on input data x.

$$\mathbb{R}^D \rightarrow F : x \rightarrow \phi(x) \tag{3}$$

Kernel PCA is an extension of PCA in nonlinear space, capable of capturing the higher-order statistics of data. Input is transformed into a new feature space, where the discrimination power of each instant from others is very high. Convert data in $\mathbb{R}^D$ to feature space F by applying a nonlinear transformation ϕ on input data x.

$$K_{ij} = exp(\frac{-1}{2\sigma^2}||x_i - x_j||) \tag{4}$$

where K_{ij} is each element in kernel matrix K, the value of σ depend on the dataset and x_i, x_j are each pixel values in the image.

The data is centered using the Eq. (5)

$$K_c = K - 1_N K - K 1_N + 1_N K 1_N \tag{5}$$

where $1_N = \frac{1}{N} * ones(N, N)$, K_c is centered kernel and N the total number of samples(pixels) in hyperspectral image. ones(m,n) means a $m \times n$ matrix having all the values are one. Similar to linear PCA, kernel PCA follows eigendecomposition on the kernel matrix K_c and finds k eigenvectors with the largest eigenvalues. Eigen decomposition of $N \times N$ matrix is computationally complex when compared with linear PCA decomposition. Thus a group of representative pixel selection from each class of pixels reduces the computational complexity [5].

The step by step procedure of Hybrid dimensionality reduction technique is shown in Algorithm 1.

Algorithm 1. Hybrid Dimensionality Reduction Technique

Input: Dataset $X \in \mathbb{R}^{N \times D}$, D is dimension of input and N is number of samples
Output: Low dimensional dataset $Y \in \mathbb{R}^{N \times k}$, k is desired dimension
1: Choose an intermediate dimension k_1 for Random projection(found $k_1 \approx (D/2)$ provide maximum efficiency)
2: Create the random matrix $R^{D \times k_1}$ using GRP and SRP
3: Perform random projection and create intermediate data $Z^{N \times k_1}$
4: **for** $i = 2$ to 30 **do**
5:     Perform nonlinear DR techniques KPCA to reduce data as $Y^{N \times i}$
6:     Find percentage of cumulative eigen value for each i
7: **end for**
8: choose $k = min(i)$, which having cumulative eigen value ≥ 95
9: **return** $Y^{N \times k}$

There are different nonlinear dimensionality reduction techniques are available in the literature. However, we choose KPCA for the proposed method. The kernel based technique is better than that of normal techniques and it helps to extract largest difference in projection points. Thus the descrimination of features will be more easier. Since KPCA using Gaussian kernel, it is good for natural images.

3   Experiments

The proposed method is evaluated by experiments in universally available hyperspectral image datasets. All executions are done on Intel(R) Xeon(R) Silver 4114 CPU @ 2.24 GHz with a RAM of 196 GB under CentOS Linux release 7.4.1708 (Core) using python3 programming implementation. Three hyperspectral datasets are used to evaluate the performance of the proposed method and state of the art techniques. They are Indian pines (IP), Pavia University (PU) and Salinas (SA) collected from the website http://www.ehu.eus/ccwint/index.php (Table 1).

Table 1. Ground truth image of various datasets and detailed description of various classes

Indian Pines (IP)		Pavia University (PU)		Salinas (SA)	
Bands	200	Bands	103	Bands	204
Spatial Dimension	145×145	Spatial Dimension	610×340	Spatial Dimension	512×214
Class	Samples	Class	Samples	Class	Samples
Background	10776	Background	164624	Background	56975
Alfalfa	46	Asphalt	6631	Brocoli_green_weeds_1	2009
Corn-notill	1428	Meadows	18649	Brocoli_green_weeds_2	3726
Corn-mintill	830	Gravel	2099	Fallow	1976
Corn	237	Trees	3064	Fallow_rough_plow	1394
Grass-pasture	483	Painted metal sheets	1345	Fallow_smooth	2678
Grass-trees	730	Bare Soil	5029	Stubble	3959
Grass-pasture-mowed	28	Bitumen	1330	Celery	3579
Hay-windrowed	478	Self-Blocking Bricks	3682	Grapes_untrained	11271
Oats	20	Shadows	947	Soil_vinyard_develop	6203
Soybean-notill	972			Corn_senesced_green_weeds	3278
Soybean-mintill	2455			Lettuce_romaine_4wk	1068
Soybean-clean	593			Lettuce_romaine_5wk	1927
Wheat	205			Lettuce_romaine_6wk	916
Woods	1265			Lettuce_romaine_7wk	1070
Buildings-Grass-Trees-Drives	386			Vinyard_untrained	7268
Stone-Steel-Towers	93			Vinyard_vertical_trellis	1807
Total Samples	21025	Total Samples	207400	Total Samples	111104

3.1 Performance Evaluation Measures

This section describes three evaluation measures, namely overall accuracy (OA), area under ROC curve (AUC), and execution time (ET) used for comparing the performance of hybrid dimensionality reduction techniques with classic nonlinear technique. The detailed explanation of evaluation measures as follows:

Overall Accuracy. Overall accuracy is a measure that states how much samples correctly mapped to its corresponding class. It is the most comfortable measure to find the performance of a classifier. Consider a dataset that has N number of samples and C class labels, then the confusion matrix M of classification is a square matrix of size $C \times C$. Overall accuracy is measured using the equation

$$OA = \frac{\sum_{i=1}^{C} M_{ii}}{N} \tag{6}$$

Diagonal elements M_{ii} in the confusion matrix is the number of samples correctly classified, and the overall accuracy of a classifier always measured on a percentage scale.

Area Under ROC Curve. The area under the ROC curve or precisely area under the curve in classification performance evaluation is an aggregate measure over all possible classification thresholds. It was used when the model ranks random positives than random negatives. It measures how well the prediction ranked than its absolute value; therefore, AUC is scale-invariant. AUC value ranges from 0 to 1.

Execution Time. The time taken for the classification of different dimensional data is measured in seconds. This paper mainly focuses on reducing the computational complexity of nonlinear DR techniques. Thus ET places a significant role in this performance evaluation.

3.2 Experiment Setup

As described in Sect. 2, the proposed algorithm is a hybrid band extraction method. The proposed method first performs Random projection on the input datasets to reduce the dimension from D to k and then perform various nonlinear dimensionality reduction techniques in this k dimensional space.

In Random projection, the projection matrix may be either dense or sparse. According to JL lemma [7], the desired dimension depends on the number of samples. If we consider the datasets mentioned above, $log(N)$ value for all the datasets are higher than their original dimension D. Thus choosing the desired dimension for random projection using this concept is not worth it. Here we choose the value of k as $D/2$ and represent the data using k Gaussian mixture elements. Compare the hybrid model using both dense and sparse projection results. The dimensionally reduced image is fed into a classifier, and the performance of the proposed method is evaluated using overall classification accuracy and execution time taken for classification. Here, parameters are assessed using the k-nearest neighbor classifier. Classification accuracy and area under the curve are calculated using a 10-fold cross-validation technique. Entire data is divided into 10 blocks, and classification is repeated 10 times by considering one block as testing data and the remaining nine blocks as training at a time.

Table 2. Percentage of cumulative Eigen values of principal components in proposed method (GRP+KPCA)

No of PCs	Percentage of Cumulative Eigenvalues		
	IP Dataset	PU Dataset	SA Dataset
2	63.18	72.14	74.15
4	73.74	77.23	79.23
6	81.54	82.12	82.23
8	88.62	87.13	88.24
10	91.97	93.56	93.89
12	94.24	94.23	94.98
14	94.86	**95.39**	**95.69**
16	**95.92**	97.06	96.78
18	97.60	98.19	98.12
20	98.15	98.47	99.01
22	98.41	99.12	99.34
24	98.88	99.38	99.56
26	98.91	99.56	99.62
28	98.99	99.61	99.67
30	99.01	99.72	99.69

4   Result Analysis and Discussion

The proposed method is evaluated using various parameters. Here our main intention is to reduce the execution time of nonlinear dimensionality reduction technique. The desired number of bands calculated using cumulative eigenvalues of the KPCA matrix. The percentage of cumulative eigenvalue for three datasets is tabulated in Table 2.

The value of overall accuracy, execution time, and area under the curve for the Indian pines dataset is given in Table 3. While comparing the classification accuracy of sparse random projection-based dimensionality reduction technique with other techniques, we found that it is very less. The execution time of the hybrid method is less than that of the KPCA technique and OA, AUC values are slightly more. Similar results got for both PU and SA datasets, and the execution time is decreased for the hybrid model.

Table 3. Evaluation parameters for Indian pines dataset in different number of bands

No of bands	OA			AUC			ET		
	KPCA	GRP+KPCA	SRP+KPCA	KPCA	GRP+KPCA	SRP+KPCA	KPCA	GRP+KPCA	SRP+KPCA
2	55.9465	55.67	55.2278	0.51	0.52	0.52	55.9465	34.447151	21. 107796
4	60.8644	59.74195	54.9736	0.59	0.59	0.54	60.8644	30.84164	21. 364091
6	64.437	64.97278	53.371	0.62	0.63	0.52	64.437	34.380761	25. 782259
8	65.0258	66.69592	57.9078	0.62	0.64	0.55	65.0258	32.620458	27. 458711
10	75.3133	67.67597	58.0792	0.64	0.64	0.55	67.3133	35.198763	28. 390743
12	78.0652	70.45697	60.8799	0.68	0.69	0.6	70.0652	40.823588	31. 904233
14	85.1402	75.72344	61.453	0.70	0.69	0.61	75.1402	42.613695	32. 64280
16	85.9887	78.14216	62.3965	0.71	0.75	0.61	78.9887	44.611529	35. 537793
18	85.1601	81.52695	65.876	0.71	0.77	0.62	80.1601	47.443079	38. 793931
20	85.2031	82.64078	65.066	0.72	0.78	0.62	82.2031	48.19298	44. 548063
22	85.7923	82.94505	65.4757	0.71	0.83	0.63	85.7923	50.539361	45. 176983
24	85.9353	82.95007	65.6996	0.72	0.83	0.61	85.9353	53.937472	46. 476957
26	85.9543	83.1339	66.5432	0.72	0.85	0.62	85.9543	56.231212	47. 122052
28	85.997	83.33042	65.0894	0.73	0.83	0.63	85.997	61.17097	49. 55423
30	86.0874	83.65406	66.8331	0.73	0.83	0.75	86.0874	62.719966	52. 883703

From Table 2, we found that the desired number of dimensions needed for classification is 16, 14, and 14 for IP, PU, and SA dataset. Table 4 compare various dimensionality reduction methods for three datasets with the desired number of bands. For the IP dataset, the overall accuracy is better in KPCA technique than the hybrid methods, whereas AUC and ET are better for hybrid techniques. The classification map for three datasets is shown in Figs. 1, 2, and 3. When considering the trade-off between classification accuracy and execution time, this hybrid method is worth maximum classification accuracy in minimum execution time.

Table 4. Summary of the result of various dimensionality reduction techniques applied in three datasets

Dataset	Method	No of bands	OA	AUC	ET (sec)
IP	KPCA	22	85.7923	0.71	85.7923
	GRP+KPCA	16	78.14216	0.75	44.61153
	SRP+KPCA	20	65.066	0.62	44.548063
PU	KPCA	20	73.8924	0.77	75.1402
	GRP+KPCA	14	88.3081	0.81	50.53936
	SRP+KPCA	20	71.1328	0.73	40.82359
SA	KPCA	16	80.549	0.82	85.9543
	GRP+KPCA	14	83.1673	0.83	56.23121
	SRP+KPCA	16	76.2298	0.81	48.19298

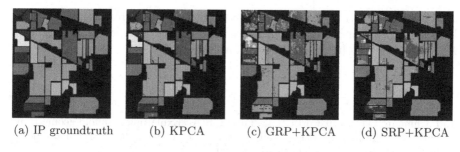

(a) IP groundtruth   (b) KPCA   (c) GRP+KPCA   (d) SRP+KPCA

Fig. 1. Classification map for Indian pines dataset

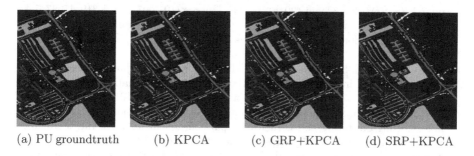

(a) PU groundtruth   (b) KPCA   (c) GRP+KPCA   (d) SRP+KPCA

Fig. 2. Classification map for Pavia University dataset

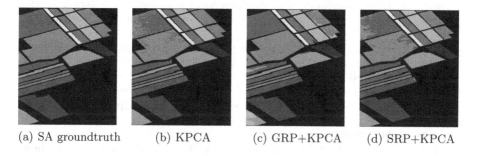

(a) SA groundtruth   (b) KPCA   (c) GRP+KPCA   (d) SRP+KPCA

Fig. 3. Classification map for Salinas dataset

5   Conclusion

This paper presented a hybrid dimensionality reduction technique for hyperspectral image analysis. Classic nonlinear dimensionality reduction techniques are computationally complex. This hybrid model initially applied the Random projection technique on the input dataset and reduced the input dimension. Sparse random projection helped to reduce the execution time due to less number of nonzero elements in the projection matrix. These band reduction outputs fed into nonlinear methods. Experiments were done on four hyperspectral datasets using GRP and SRP based hybrid models. Experimental results demonstrate that these hybrid methods reduce the computational complexity of nonlinear DR techniques with better classification accuracy.

References

1. Bachmann, C.M., Ainsworth, T.L., Fusina, R.A.: Exploiting manifold geometry in hyperspectral imagery. IEEE Trans. Geosci. Remote Sens. **43**(3), 441–454 (2005). https://doi.org/10.1109/TGRS.2004.842292

2. Bruzzone, L., Liu, S., Bovolo, F., Du, P.: Change detection in multitemporal hyperspectral images. In: Ban, Y. (ed.) Multitemporal Remote Sensing. RSDIP, vol. 20, pp. 63–88. Springer, Cham (2016). https://doi.org/10.1007/978-3-319-47037-5_4

3. Dasgupta, S., Gupta, A.: An elementary proof of a theorem of Johnson and Lindenstrauss. Random Struct. Algorithms **22**(1), 60–65 (2003). https://doi.org/10.1002/rsa.10073

4. Datta, A., Ghosh, S., Ghosh, A.: Band elimination of hyperspectral imagery using partitioned band image correlation and capacitory discrimination. Int. J. Remote Sens. **35**(2), 554–577 (2014). https://doi.org/10.1080/01431161.2013.871392

5. Datta, A., Ghosh, S., Ghosh, A.: Unsupervised band extraction for hyperspectral images using clustering and kernel principal component analysis. Int. J. Remote Sens. **38**(3), 850–873 (2017). https://doi.org/10.1080/01431161.2016.1271470

6. Jensen, J.: Remote Sensing of the Environment: An Earth Resource Perspective, 2nd edn. Prentice Hall (2007)

7. Johnson, W., Lindenstrauss, J.: Extensions of Lipschitz mappings into a Hilbert space. In: Conference in Modern Analysis and Probability (New Haven, Conn., 1982), Contemporary Mathematics, vol. 26, pp. 189–206. American Mathematical Society (1984)

8. Li, D., Wang, X., Cheng, Y.: Spatial-spectral neighbour graph for dimensionality reduction of hyperspectral image classification. Int. J. Remote Sens. 1–23 (2019). https://doi.org/10.1080/01431161.2018.1562587

9. Luo, G., Chen, G., Tian, L., Qin, K., Qian, S.E.: Minimum noise fraction versus principal component analysis as a preprocessing step for hyperspectral imagery denoising. Can. J. Remote Sens. **42**(2), 106–116 (2016). https://doi.org/10.1080/07038992.2016.1160772

10. Martínez, A.M., Kak, A.C.: PCA versus LDA. IEEE Trans. Pattern Anal. Mach. Intell. **23**(2), 228–233 (2001). https://doi.org/10.1109/34.908974

11. Prasad, S., Bruce, L.M.: Limitations of principal components analysis for hyperspectral target recognition. IEEE Geosci. Remote Sensing Lett. **5**(4), 625–629 (2008)

12. Roweis, S.T., Saul, L.K.: Nonlinear dimensionality reduction by locally linear embedding. Science **290**(5500), 2323–2326 (2000). https://doi.org/10.1126/science.290.5500.2323

13. Liu, S., Du, Q., Tong, X.: Band selection for change detection from hyperspectral images (2017). https://doi.org/10.1117/12.2263024

14. Thenkabail, P.S., Smith, R.B., De Pauw, E.: Hyperspectral vegetation indices for determining agricultural crop characteristics. Remote Sens. Environ. 1–37 (2000)

15. Theodoridis, S., Koutroumbas, K.: Pattern Recognition, 3rd edn. Academic Press Inc., Orlando (2006)

16. Venkatasubramanian, S., Wang, Q.: The Johnson-Lindenstrauss transform: an empirical study. In: Proceedings of the Meeting on Algorithm Engineering & Experiments, ALENEX 2011, pp. 164–173. Society for Industrial and Applied Mathematics, Philadelphia (2011). http://dl.acm.org/citation.cfm?id=2790248.2790264
17. Wang, J., Chang, C.I.: Independent component analysis-based dimensionality reduction with applications in hyperspectral image analysis. IEEE Trans. Geosci. Remote Sens. **44**(6), 1586–1600 (2006). https://doi.org/10.1109/TGRS.2005.863297

Clustering Spatial Data via Mixture Models with Dynamic Weights

Allou Samé[⊠]

COSYS-GRETTIA, Univ Gustave Eiffel, IFSTTAR, F-77447 Cedex 2
Marne-la-Vallée, France
allou.same@ifsttar.fr

Abstract. This article introduces a spatial mixture model for the modeling and clustering of georeferenced data. In this model, the spatial dependence of the data is taken into account through the mixture weights, which are modeled by logistic transformations of spatial coordinates. In this way, the observations are supposed to be independent but not identically distributed, their dependence being transferred to the parameters of these logistic functions. A specific EM algorithm is used for parameter estimation via the maximum likelihood method, which incorporates a Newton-Raphson algorithm for estimating the logistic functions coefficients. The experiments, carried out on synthetic images, give encouraging results in terms of segmentation accuracy.

Keywords: Clustering · Segmentation · Spatial data · Mixture model · Hidden logistic random field · EM algorithm

1 Introduction

In many applicative areas, it is often required to analyse spatially organized data in order to extract some relevant information. Segmentation, which consists in partitioning such data into homogeneous regions, is one of the fundamental methods used. The Gaussian mixture-model-based clustering approach [10], implemented via the Expectation-Maximization (EM) algorithm [6] constitutes a reference method in this framework, but which is designed for the clustering of multivariate data which are not necessarily spatial in nature. Among the statistical approaches allowing the spatial aspect of the data to be effectively taken into account, methods based on hidden Markov random fields can be seen as a reference [1,2,4]. Other approaches which consist in modifying the usual Gaussian mixture model have also been proposed to solve the problem. In [3], a spatial prior distribution on the mixture weights and the parameters are estimated in the Bayesian framework via an EM algorithm. To address the so-called perceptual grouping problem, the Neural EM algorithm has been introduced in [8], where the parameters of the conditional probability distributions are formalized as neural networks. In [15], an analog problem was addressed by modeling the mixture weights using neural networks, which explicitly take into account the data spatial correlations. In [14], a spatial mixture model is proposed within a

© Springer Nature Switzerland AG 2020
W. Lu and K. Q. Zhu (Eds.): PAKDD 2020 Workshops, LNAI 12237, pp. 128–138, 2020.
https://doi.org/10.1007/978-3-030-60470-7_13

Bayesian framework. The spatial regularization parameter involved in this model is adaptively estimated from the data. To find regions suitable for the implantation of solar energy, a clustering method has been proposed in [13] for spatially distributed functional data. This method consists in locally averaging the curves in the regions defined by a randomly generated Voronoï diagram constructed from spatial coordinates, reducing the dimension of the curves by a principal component analysis, perform clustering on the resulting low-dimensional data, and then matching the cluster labels among the bootstrap Voronoï diagrams.

In the same line as the methods based on spatial mixture distributions, this article proposes an extension of spatial mixture models for georeferenced data. The data spatial dependencies are taken into account through the mixture weights, which are modeled by logistic transformations of linear or nonlinear functions of the spatial coordinates. In this way, the observations are supposed to be independent but not identically distributed, their dependencies having been transferred to the parameters of these logistic functions. This model constitutes the spatial extension of the mixture model dedicated to the segmentation of time series, proposed in [5,11]. The article is organized as follows: Sect. 2 describes the proposed model and its parameter estimation algorithm. Section 3 evaluates this algorithm within the framework of image segmentation. Some perspectives of this work are given in the last section.

2   Hidden Logistic Random Field Mixture Model

In the rest of the article, the spatial data to be partitioned will be denoted by $(x_1, \ldots, x_n)$, where $x_i \in \mathbb{R}^d$ will be associated to a geographic site s_i ($i = 1, \ldots, n$). In the area of geographic sciences, for example, the statistical unit x_i generally represents a vector of demographic, environmental or socio-economic characteristics associated with a parcel, a city or a country, and the site s_i corresponds to the geographic coordinates (latitude, longitude) of x_i. In the field of image processing, the statistical units are the pixels, characterized by attributes such as the gray level or the intensity of the Red/Green/Blue colors. These pixels are identified by their coordinates in the image. We will assume, to simplify, that s_i is defined by the pair of non-random spatial coordinates $(u_i, v_i) \in \mathbb{R}^2$. The model that we propose makes the assumption, as for the classical Gaussian mixture models [9,10], that each observation is distributed according to a mixture of K Gaussian densities but whose proportions are specific functions of the spatial coordinates (u_i, v_i). This mixture model is defined by:

$$p(x_i; \Phi) = \sum_{k=1}^{K} \pi_k(u_i, v_i; \alpha) \mathcal{N}(x_i; \mu_k, \Sigma_k), \tag{1}$$

where the μ_k and Σ_k are the means and covariance matrices of the Gaussian components, which are defined in the space $\mathbb{R}^d$, and $\Phi = \{\alpha, (\mu_k, \Sigma_k)_{k=1,\ldots,K}\}$ denotes the set of parameters of the model. To take into account the spatial regularity of the data in the classification, we define the probability $\pi_k(u_i, v_i; \alpha)$

by the following logistic transformation of a polynomial function of the spatial coordinates:

$$\pi_k(u_i, v_i; \boldsymbol{\alpha}) = \frac{\exp(\boldsymbol{r}(u_i, v_i)^T \boldsymbol{\alpha}_k)}{\sum_{\ell=1}^{K} \exp(\boldsymbol{r}(u_i, v_i)^T \boldsymbol{\alpha}_\ell)}, \tag{2}$$

where $\boldsymbol{\alpha} = (\boldsymbol{\alpha}_1^T, \ldots, \boldsymbol{\alpha}_K^T)^T$ is the set of coefficients of the logistic functions, $\boldsymbol{r}(u_i, v_i) \in \mathbb{R}^q$ is the vector of monomial associated to a polynomial function of the couple (u_i, v_i), and $\boldsymbol{\alpha}_k \in \mathbb{R}^q$ denotes its coefficients vector. Table 1 gives examples of vectors $\boldsymbol{r}(u_i, v_i)$ for different polynomial degrees. It can be observed that the logistic functions defined by Eq. (1) satisfy the constraints $\sum_{k=1}^{K} \pi_k(u_i, v_i; \boldsymbol{\alpha}) = 1$ et $0 < \pi_k(u_i, v_i; \boldsymbol{\alpha}) < 1$.

Table 1. Vectors $\boldsymbol{r}(u_i, v_i)$ associated to different polynomial orders

Order 1	$q = 3$	$\boldsymbol{r}(u_i, v_i) = (1, u_i, v_i)^T$
Order 2	$q = 6$	$\boldsymbol{r}(u_i, v_i) = (1, u_i, v_i, u_i^2, u_i v_i, v_i^2)^T$
Order 3	$q = 10$	$\boldsymbol{r}(u_i, v_i) = (1, u_i, v_i, u_i^2, u_i v_i, v_i^2, u_i^3, u_i^2 j, u_i v_i^2, v_i^3)^T$

The identifiability of the model, that is to say the equivalence $p(\boldsymbol{x}_i; \boldsymbol{\Phi}_1) = p(\boldsymbol{x}_i; \boldsymbol{\Phi}_2) \iff \boldsymbol{\Phi}_1 = \boldsymbol{\Phi}_2$, is ensured by setting the last parameter of the logistic functions $(\boldsymbol{\alpha}_K)$ to the null vector (Jiang and Tanner, 1999). Under these conditions, which will be assumed in the rest of this article, the vector $\boldsymbol{\alpha}$ therefore belongs to the space $\mathbb{R}^{q(K-1)}$.

Note here that our model can also be posed as a generative model with latent variables, the latent variables $(z_1, \ldots, z_n)$ being generated independently according to the multinomial law $\mathcal{M}(1; \pi_1(u_i, v_i), \ldots, \pi_K(u_i, v_i))$ such that $p(z_i; \boldsymbol{\Phi}) = \pi_{z_i}(u_i, v_i)$, and the observations $(\boldsymbol{x}_1, \ldots, \boldsymbol{x}_n)$ being generated, conditionally on $(z_1, \ldots, z_n)$, according to the Gaussian density $p(\boldsymbol{x}_i | z_i; \boldsymbol{\Phi}) = \mathcal{N}(\boldsymbol{x}_i; \boldsymbol{\mu}_{z_i}, \boldsymbol{\Sigma}_{z_i})$. Figure 1 shows two color images generated according to the proposed model and their corresponding logistic mixture weights for polynomial orders 1 and 2, whose parameters are given in Table 2. We note in particular that the simulated data respect the spatial constraints imposed by the logistic probabilities.

Table 2. Simulated parameters for two data configurations

	Situation 1		Situation 2	
α_1	$(16.27 \;-0.47 \; 0.23)^T$		$(-55.00 \; 1.56 \; 0.19 \; -0.0117 \; 0.0002 \; -0.0014)^T$	
α_2	$(-14.03 \; -0.17 \; 0.53)^T$		$(-91.00 \; 0.32 \; 2.42 \; -0.0027 \; 0.0008 \; -0.0169)^T$	
α_3	$(0 \; 0 \; 0)^T$		$(0 \; 0 \; 0 \; 0 \; 0 \; 0)^T$	
$\mu_1 \; \Sigma_1$	$(255 \; 0 \; 0)^T$	$100 \times \mathbf{I}_{3,3}$	$(255 \; 0 \; 0)^T$	$100 \times \mathbf{I}_{3,3}$
$\mu_2 \; \Sigma_2$	$(0 \; 255 \; 0)^T$	$100 \times \mathbf{I}_{3,3}$	$(0 \; 255 \; 0)^T$	$100 \times \mathbf{I}_{3,3}$
$\mu_3 \; \Sigma_3$	$(0 \; 0 \; 255)^T$	$100 \times \mathbf{I}_{3,3}$	$(0 \; 0 \; 255)^T$	$100 \times \mathbf{I}_{3,3}$

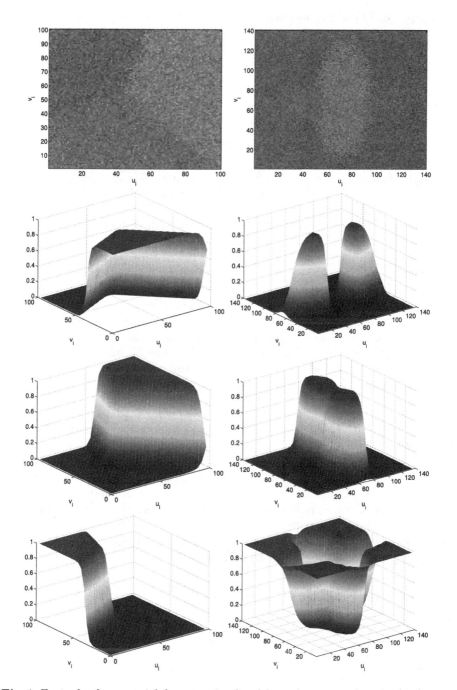

Fig. 1. Example of two spatial data sets simulated from the proposed model (top) with $K = 3$ classes, and corresponding mixture weights (bottom)

2.1   Spatial Data Segmentation

The proposed model leads to a segmentation $(\Omega_1, \ldots, \Omega_K)$, where the set Ω_k, which is obtained by maximizing logistic probabilities, is defined by

$$
\begin{aligned}
\Omega_k &= \left\{ (u,v) : \pi_k(u,v;\boldsymbol{\alpha}) = \max_{1 \le \ell \le K} \pi_\ell(u,v;\boldsymbol{\alpha}) \right\} \\
&= \bigcap_{\ell=1}^{K} \left\{ (u,v) : \log(\pi_k(u,v;\boldsymbol{\alpha})/\pi_\ell(u,v;\boldsymbol{\alpha})) \ge 0 \right\} \\
&= \bigcap_{\ell=1}^{K} \left\{ (u,v) : (\boldsymbol{\alpha}_k - \boldsymbol{\alpha}_\ell)^T \, \boldsymbol{r}(u,v) \ge 0 \right\}.
\end{aligned}
\tag{3}
$$

Consequently, in the particular case where the mixture proportions are logistic transformations of first order polynomials of (u, v), the region Ω_k is convex as the intersection of convex parts of $\mathbb{R}^2$. Note that posterior probabilities can also be used to partition the data if strict geometric constraints such as convexity are not imposed on the segments.

It should be emphasized that, the model described above, which will be called the "Hidden Logistic Random Field (HLRF)" model, appears especially appropriate for problems where each class, in the feature space, is associated to one region of the image. The more the shape of the segments will be complex, the more a higher polynomial order will be required by the logistic function for the segmentation to be sufficiently accurate. This kind of image segmentation approach is often called a "one-class-segment" method. We propose to extend the HLRF concept to address problems in which a class in the feature space can appear in geographically separated regions of an image. This second version of the HLRF model is based on the following mixture weights:

$$
\pi_k(u_i, v_i; \boldsymbol{\alpha}, \boldsymbol{\beta}) = \frac{\exp(\boldsymbol{r}(u_i,v_i)^T \boldsymbol{\alpha}_k + \boldsymbol{w}_{\sim i}^T \boldsymbol{\beta}_k)}{\sum_{\ell=1}^{K} \exp(\boldsymbol{r}(u_i,v_i)^T \boldsymbol{\alpha}_\ell + \boldsymbol{w}_{\sim i}^T \boldsymbol{\beta}_\ell)},
\tag{4}
$$

where $\boldsymbol{\beta} = (\boldsymbol{\beta}_1, \ldots, \boldsymbol{\beta}_K)$ and $\boldsymbol{w}_{\sim i}$ is the average value of the observations belonging to neighbors of $\boldsymbol{x}_i$ (a system of 8 neighboring observations is considered in this article). The complete set of parameters of the model is now given by $\boldsymbol{\Phi} = (\boldsymbol{\alpha}, \boldsymbol{\beta}, (\boldsymbol{\mu}_k, \boldsymbol{\Sigma}_k))$. In this setting, the prior probabilities associated to a couple $(\boldsymbol{x}_i, \boldsymbol{s}_i)$ results from a compromise between the spatial coordinates $\boldsymbol{s}_i = (u_i, v_i)$ and the attributes of the neighbors of $\boldsymbol{x}_i$. In the following, the model defined by Eqs. (1) and (2) will be called HLRF1 and the one described by Eqs. (1) and (4) will be called HLRF2.

2.2   Parameter Estimation Algorithm

This section focuses on the parameter estimation of the HLRF2 model, the HLRF1 model being a particular case of this one. We naturally propose to maximize the log-likelihood criterion defined by

$$
\mathcal{L}(\boldsymbol{\Phi}) = \log \prod_{i=1}^{n} p(\boldsymbol{x}_{ij}; \boldsymbol{\Phi}) = \sum_i \log \sum_k \pi_k(u_i, v_i; \boldsymbol{\alpha}, \boldsymbol{\beta}) \, \mathcal{N}(\boldsymbol{x}_i; \boldsymbol{\mu}_k, \boldsymbol{\Sigma}_k).
\tag{5}
$$

using a dedicated EM algorithm, which consists in starting from an initial parameter $\boldsymbol{\Phi}^{(0)}$ and computing iteratively the parameter $\boldsymbol{\Phi}^{(c+1)}$ which maximizes the auxiliary function

$$\mathcal{F}^{(c)}(\boldsymbol{\Phi}) = \sum_{i,k} \tau_{ik}^{(c)} \log(\pi_k(u_i, v_i; \boldsymbol{\alpha}, \boldsymbol{\beta}) \mathcal{N}(\boldsymbol{x}_i; \boldsymbol{\mu}_k, \boldsymbol{\Sigma}_k)), \tag{6}$$

with

$$\tau_{ik}^{(c)} = \frac{\exp(\boldsymbol{r}(u_i, v_i)^T \boldsymbol{\alpha}_k^{(c)} + \boldsymbol{w}_{\sim i}^T \boldsymbol{\beta}_k^{(c)}) \mathcal{N}(\boldsymbol{x}_i; \boldsymbol{\mu}_k^{(c)}, \boldsymbol{\Sigma}_k^{(c)})}{\sum_{\ell=1}^{K} \exp(\boldsymbol{r}(u_i, v_i)^T \boldsymbol{\alpha}_\ell^{(c)} + \boldsymbol{w}_{\sim i}^T \boldsymbol{\beta}_\ell^{(c)}) \mathcal{N}(\boldsymbol{x}_i; \boldsymbol{\mu}_\ell^{(c)}, \boldsymbol{\Sigma}_\ell^{(c)})}. \tag{7}$$

This procedure is known to converge towards a local maximum of $\mathcal{L}$. The maximization of $\mathcal{F}^{(c)}$ with respect to $\boldsymbol{\mu}_k$ and $\boldsymbol{\Sigma}_k$ is analogous to that of the usual Gaussian mixture model, and the updates $\boldsymbol{\alpha}^{(c+1)}$ and $\boldsymbol{\beta}^{(c+1)}$ of the logistic functions are, for their part, obtained by solving a convex logistic regression problem. This problem can efficiently be solved by a multiclass Iteratively Reweighted Least Squares (IRLS) Algorithm [7] which is equivalent to a Newton-Raphson algorithm. By denoting the first and second derivatives of $\mathcal{F}^{(c)}$ w.r.t. $\boldsymbol{\lambda} = (\boldsymbol{\alpha}^T, \boldsymbol{\beta}^T)^T$ as

$$\mathbf{g}(\boldsymbol{\alpha}, \boldsymbol{\beta}) = \big(\mathbf{g}_k(\boldsymbol{\alpha}, \boldsymbol{\beta})\big)_{1 \le k \le K-1} \tag{8}$$

$$\mathbf{H}(\boldsymbol{\alpha}, \boldsymbol{\beta}) = \big(\mathbf{H}_{k\ell}(\boldsymbol{\alpha}, \boldsymbol{\beta})\big)_{1 \le k, \ell \le K-1}, \tag{9}$$

where

$$\mathbf{g}_k = \frac{\partial \mathcal{F}^{(c)}}{\partial \boldsymbol{\lambda}_k} \quad \mathbf{H}_{k\ell} = \frac{\partial^2 \mathcal{F}^{(c)}}{\partial \boldsymbol{\lambda}_k \partial \boldsymbol{\lambda}_\ell^T}, \tag{10}$$

with $\delta_{k\ell} = 1$ if $k = \ell$ and $\delta_{k\ell} = 0$ otherwise, the main steps of the IRLS algorithm can be summarized by Algorithm 1. A synthesis of the EM-type iterative procedure aiming at maximizing $\mathcal{L}$ is given by Algorithm 2.

Algorithm 1: IRLS

Input: posterior probabilities $(\tau_{ik}^{(c)})$, threshold ε
$(\boldsymbol{\alpha}, \boldsymbol{\beta}) \leftarrow \mathbf{0}$
repeat

$\quad \mathbf{g}_k \quad \leftarrow \sum_i \tau_{ik}^{(c)} - \pi_k(u_i, v_i; \boldsymbol{\alpha}, \boldsymbol{\beta})$
$\quad \mathbf{H}_{k\ell} \leftarrow -\sum_i \pi_k(u_i, v_i; \boldsymbol{\alpha}, \boldsymbol{\beta}) \times (\delta_{k\ell} - \pi_\ell(u_i, v_i; \boldsymbol{\alpha}, \boldsymbol{\beta}))$
$\quad \begin{pmatrix} \boldsymbol{\alpha} \\ \boldsymbol{\beta} \end{pmatrix} \leftarrow \begin{pmatrix} \boldsymbol{\alpha} \\ \boldsymbol{\beta} \end{pmatrix} - \mathbf{H}^{-1} \mathbf{g}$

until $\|\mathbf{g}\| < \varepsilon$;
$(\boldsymbol{\alpha}^{(c+1)}, \boldsymbol{\beta}^{(c+1)}) \leftarrow (\boldsymbol{\alpha}, \boldsymbol{\beta})$
Output: logistic functions parameters $(\boldsymbol{\alpha}^{(c+1)}, \boldsymbol{\beta}^{(c+1)})$

Algorithm 2: EM

Input: data $(x_i)_{i=1,\ldots,n}$, number or segments K, polynomial order q, threshold
$\quad\quad \varepsilon$, initial parameter $\boldsymbol{\Phi}^{(0)}$

$c \leftarrow 0$

repeat

$\quad$ *E-step* Compute the posterior probabilities:

$$\tau_{ik}^{(c)} \leftarrow \frac{\exp\left(r(u_i,v_i)^T \alpha_k^{(c)} + w_{\sim i}^T \beta_k^{(c)}\right) \mathcal{N}(x_i; \mu_k^{(c)}, \Sigma_k^{(c)})}{\sum_{\ell=1}^K \exp\left(r(u_i,v_i)^T \alpha_\ell^{(c)} + w_{\sim i}^T \beta_\ell^{(c)}\right) \mathcal{N}(x_i; \mu_\ell^{(c)}, \Sigma_\ell^{(c)})}$$

$\quad$ *M-step* Update the parameters:

$$\begin{pmatrix} \alpha^{(c+1)} \\ \beta^{(c+1)} \end{pmatrix} \leftarrow \text{IRLS}\left(\left(\tau_{ik}^{(c)}\right)\right)$$

$$\mu_k^{(c+1)} \leftarrow \left(\sum_i \tau_{ik}^{(c)} x_i\right) / \left(\sum_i \tau_{ik}^{(c)}\right)$$

$$\Sigma_k^{(c+1)} \leftarrow \frac{\sum_i \tau_{ik}^{(c)} \left(x_i - \mu_k^{(c+1)}\right)\left(x_i - \mu_k^{(c+1)}\right)^T}{\sum_i \tau_{ik}^{(c)}}$$

$\quad c \leftarrow c+1$

until *Change in* $\mathcal{L} < \varepsilon$;

Output: Parameter vector $\widehat{\boldsymbol{\Phi}}$

---

3   Experiments

In this section, the proposed method has been evaluated using synthetic data. Let us recall that in the case of images, the spatial coordinates (u_i, v_i) belong to a regular two-dimensional grid $\{1,\ldots,U\} \times \{1,\ldots,V\}$, where U and V are the dimensions of the image. We assume in this article that the observations x_i are vectors of $\mathbb{R}^3$ indicating the intensity of the colors Red/Green/Blue.

The four following methods are compared:

- GMM: the classical Expectation Maximization algorithm associated to the Gaussian mixture model [9];
- MRF: the algorithm based on the Potts hidden Markov random field implemented via a modified version of the EM algorithm called Neighborhood Expectation Maximization algorithm [1];
- HLRF1 and HLRF2: the proposed method associated to the HLRF1 and HLRF2 models.

Two configurations of synthetic colored images of size 200×200 pixels have been considered:

- the configuration 1, corresponding to an image with $K = 3$ connected segments;
- the configuration 2, corresponding to an image with $K = 2$ non connected segments.

For each configuration, we considered three different noise levels: level 1 (easy), level 2 (medium) and level 3 (difficult). Figure 2 displays the true segmentation corresponding to these configurations, and Fig. 3 shows, for each situation, the images generated according to these three noise levels.

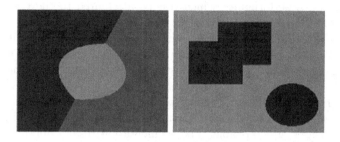

Fig. 2. True segmentation for configuration 1 (left) and configuration 2 (right)

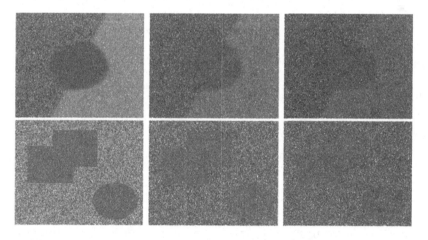

Fig. 3. Images corresponding to configuration 1 (top) and configuration 2 (bottom) for noise levels 1 (left), 2 (center) and 3 (right)

The four algorithms were run with the true numbers of clusters ($K = 3$ and $K = 2$). For the MRF approach, the spatial coefficient which determines the smoothness of the partition has been varied from 0.1 to 4 by step of 0.1, and the coefficient which provides the lowest misclassification rate has been retained. HLRF1 was applied with respective polynomial orders 2 ($q = 6$) and 8 ($q = 45$) for situations 1 and 2, and HLRF2 was used with a second order polynomial ($q = 6$) for both situations. The parameters (μ_k) and (Σ_k) have been initialized by running the standard EM algorithm for Gaussian mixture models, and the logistic regression parameters (α, β) have been set to the null vector.

To assess the quality of the partitioning, we used the Normalized Mutual Information (NMI), and the Adjusted Rand Index (ARI). Tables 3 and 4 display these criteria for the four compared approaches and for the different noise levels. For situation 1, not surprisingly, GMM, which does not take into account the spatial aspect of the data, gives poor results. HLRF1 performs better than its competitors, a performance which can be attributed to the true structure of the segments, which, for the most part are convex or connected. The results obtained with HLRF2 are very close to those yielded by HLRF1. For its part, the GMM approach did not perform well for noise levels 2 and 3. We also observed that MRF tended to find 2 segments instead of 3 segments. For situation 2, we obtained good performances of HLRF2, especially for high noise levels. Nevertheless, MRF remains highly competitive. HLRF1 requires, in this situation, a high polynomial order of logistic function (8th order), which can make the algorithm more slower and unstable. The segmentation results obtained with the four algorithms, for the noise level 2, are displayed on Fig. 4.

Table 3. NMI criterion obtained with the four compared methods

	Configuration 1			Configuration 2		
	Noise lev. 1	Noise lev. 2	Noise lev. 3	Noise lev. 1	Noise lev. 2	Noise lev. 3
GMM	0.70	0.25	0.11	0.66	0.14	0.06
MRF	0.96	0.75	0.76	**0.99**	0.94	**0.87**
HLRF1	**0.98**	**0.98**	**0.97**	0.94	0.82	0.75
HLRF2	0.94	0.97	0.96	0.96	**0.95**	**0.87**

Table 4. ARI criterion obtained with the four compared methods

	Configuration 1			Configuration 2		
	Noise lev. 1	Noise lev. 2	Noise lev. 3	Noise lev. 1	Noise lev. 2	Noise lev. 3
GMM	0.80	0.36	0.12	0.78	0.20	0.08
MRF	0.98	0.72	0.72	**0.99**	**0.97**	0.92
HLRF1	**0.99**	**0.99**	**0.99**	0.97	0.90	0.85
HLRF2	0.97	**0.99**	0.98	**0.99**	**0.97**	**0.93**

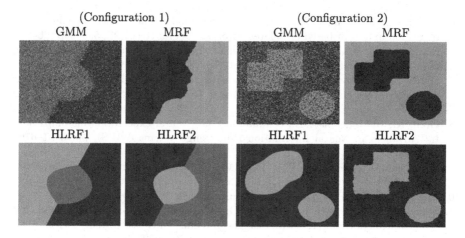

Fig. 4. Segmentations estimated for configurations 1 and 2 with noise level 2

4   Conclusion

A spatial latent variable model has been proposed in this paper for spatial data segmentation and modeling. It involves modeling georeferenced observations according to a mixture model whose proportions are logistic transformations of polynomial functions of spatial coordinates. In this way, the spatial dependence of the data can be taken into account without tuning any smoothing constant. An extension of the latter model has also been proposed to better address the issue of segments which can appear in geographically separated regions. The experiments conducted on noisy synthetic images have shown encouraging results. Some experiments dealing with real images are being investigated. The logistic functions used as the mixture weights can be made more complex by exploiting, for example, deep neural networks, which will lead to a Generalized EM algorithm. Although the issue of choosing the appropriate polynomial order and number of segments K has not been addressed in this paper, it still remains an important aspect to be considered. Basically, the Bayesian Information Criterion (BIC) [12] should be considered.

References

1. Ambroise, C., Govaert, G.: Convergence of an EM-type algorithm for spatial clustering. Pattern Recogn. Lett. **19**, 919–927 (1998)
2. Besag, J.: On the statistical analysis of dirty pictures. J. Roy. Stat. Soc. B **48**(3), 259–302 (1986)
3. Blekas, K., Likas, A., Galatsanos, N.P., Lagaris, I.E.: A spatially constrained mixture model for image segmentation. IEEE Trans. Neural Netw. **16**(2), 494–498 (2005)
4. Celeux, G., Forbes, F., Peyrard, N.: EM procedures using mean field-like approximations for Markov model-based image segmentation. Pattern Recogn. **36**(1), 131–144 (2003)

5. Chamroukhi, F., Samé, A., Govaert, G., Aknin, P.: Time series modeling by a regression approach based on a latent process. Neural Netw. **22**, 593–602 (2009)
6. Dempster, A.P., Laird, N.M., Rubin, D.B.: Maximum likelihood from incomplete data via the EM algorithm. J. Roy. Stat. Soc. B **39**, 1–38 (1977)
7. Green, P.: Iteratively reweighted least squares for maximum likelihood estimation, and some robust and resistant alternatives. J. Roy. Stat. Soc. Ser. B **46**(2), 149–192 (1984)
8. Greff, K., van Steenkiste, S., Schmidhuber, J.: Neural expectation maximization. In: Advances in Neural Information Processing Systems (NIPS), pp. 6691–6701 (2017)
9. McLachlan, G.J., Krishnan, T.: The EM Algorithm and Extensions. Wiley, New York (2008)
10. McLachlan, G.J., Peel, D.: Finite Mixture Models. Wiley, New York (2000)
11. Samé, A., Chamroukhi, F., Govaert, G., Aknin, P.: Model-based clustering and segmentation of time series with changes in regime. Adv. Data Anal. Classif. **5**(4), 301–321 (2011). https://doi.org/10.1007/s11634-011-0096-5
12. Schwarz, G.: Estimating the number of components in a finite mixture model. Ann. Stat. **6**, 461–464 (1978)
13. Secchi, P., Vantini, S., Vitelli, V.: Bagging Voronoi classifiers for clustering spatial functional data. Int. J. Appl. Earth Obs. Geoinf. **22**, 53–64 (2013). Spatial Statistics for Mapping the Environment
14. Woolrich, M.W., Behrens, T.E.J., Beckmann, C.F., Smith, S.M.: Mixture models with adaptive spatial regularization for segmentation with an application to FMRI data. IEEE Trans. Med. Imaging **24**(1), 1–11 (2005)
15. Yuan, J., Li, B., Xue, X.: Spatial mixture models with learnable deep priors for perceptual grouping. In: AAAI-2019, vol. 33, no. 1 (2019)

Ninth Workshop on Biologically Inspired Techniques for Data Mining (BDM 2020)

Temporal Convolutional Network Based Transfer Learning for Structural Health Monitoring of Composites

Shaista Hussain[1](✉), Luu Trung Pham Duong[2], Nagarajan Raghavan[2], and Mark Hyunpong Jhon[1]

[1] Institute of High Performance Computing, ASTAR, Singapore, Singapore
`hussains@ihpc.a-star.edu.sg`
[2] Singapore University of Technology and Design, Singapore, Singapore

Abstract. Composite materials have become extremely important for several engineering applications due to their superior mechanical properties. However, a major challenge in the use of composites is to detect, locate and quantify fatigue induced damage, particularly delamination, by using limited experimental data. The use of guided Lamb wave based health monitoring with embedded sensors has emerged as a potential solution to effectively predict delamination size. To do this, machine learning prediction models have been used in the past, however, a transfer learning approach which can address the problem of inadequate labeled data by allowing the use of a pretrained model for predicting damage in a new composite specimen, has not been explored in this field. This paper proposes a temporal convolutional network (TCN) based transfer learning (TCN-trans) scheme for predicting delamination damage using sensor measurements. The application of proposed framework is demonstrated on Lamb wave sensor dataset collected from fatigue experiments measuring the evolution of damage in carbon fiber reinforced polymer (CFRP) cross-ply laminates. The results show that TCN-trans yields better damage prediction by fine-tuning a pretrained model with a small number of test specimen samples as compared to a TCN trained only on the test specimen data.

Keywords: Temporal convolutional networks · Transfer learning · Composite damage prediction

1 Introduction

The use of carbon fiber reinforced polymer (CFRP) composite materials in aerospace structures has dramatically increased in the last two decades due to their superior mechanical properties, including high strength, stiffness, light weight, and excellent corrosion resistance [1]. In spite of these benefits, composites can still fail catastrophically during service. One important failure mode is mechanical fatigue, which occurs when a part is subjected to cyclic loading. The

© Springer Nature Switzerland AG 2020
W. Lu and K. Q. Zhu (Eds.): PAKDD 2020 Workshops, LNAI 12237, pp. 141–152, 2020.
https://doi.org/10.1007/978-3-030-60470-7_14

mechanism for failure is often complex, involving multiple interacting damage modes operating on multiple length scales. For example, in a composite panel, damage can initiate in the polymer matrix in the form of micro-cracks through the ply thickness. As the number of micro-cracks increase under increasing load, the composite plies may separate, leading to delamination along the ply interface and a concomitant knockdown in stiffness and strength of the panel. Damage in the panel can then lead to ultimate failure of the composite structure [2]. Although standard non-destructive evaluation (NDE) techniques such as X-ray and dye penetrant testing can be used to determine the existing level of damage in the part [3], most of these methods are not suitable for structural health monitoring (SHM) of a component in service, as they are typically slow and require disassembly and reassembling of the structural component.

In contrast, elastic wave-based methods have shown great potential in both NDE and SHM, where the commercial availability of low-cost piezoelectric transducers can enable the inspection of a large areas in a short time scale [4]. Since Lamb waves [5] interact strongly with structural defects, they are promising for damage assessment in both NDE and SHM applications. Guided Lamb waves have thus been used for diagnosis of composite structures, including locating damage [1], studying the effect of matrix micro-cracks and delamination on the velocity of Lamb wave [2], and to estimate matrix micro-crack density per path and location of delamination [6]. However, the inverse problem of identifying damage from the Lamb-wave signal remains challenging [7]. Further, the best practices for optimizing the use of transducers (for instance interrogation frequency) are not clearly defined.

Machine learning methods have been used successfully for predicting the trend, classification, and target detection, in many applications. Machine learning models like, linear regression, support vector machine and random forest, were used to predict delamination area by extracting five user-defined features from the Lamb wave signals recorded from composite material [8]. Another approach for predicting delamination area involved the use of principal component regression for extracting damage sensitive features of Lamb wave sensor signal [9]. In [10], a recurrent neural network (RNN) using three predefined inputs was proposed to predict the remaining useful life (RUL) of composites. Most of these machine learning applications for damage prediction in composite materials, use predefined features extracted from the sensor data, where the knowledge about important or relevant features has been incorporated in the machine learning model. The use of deep learning models can circumvent the need of user-defined feature extraction by performing feature self-learning from the raw input data to compute high-level representations [11]. However, the lack of sufficient experimental data limits the use of deep learning algorithms in this field. Moreover, another challenge is that these algorithms need to be retrained every time a new composite specimen is presented, which will lead to long training times.

These issues can be addressed by the use of a deep transfer learning based approach, where transfer learning algorithms transfer latent knowledge learnt from a source to a target domain [12]. Deep RNN based transfer learning was used to predict RUL of turbofan engines by first training on a source data followed by using fewer samples from target data set [13]. In this paper, we propose a temporal convolutional network based transfer learning (TCN-trans) method for prediction of delamination area in composite materials. The main contributions of this paper include:

(1) We have used deep learning for damage prediction in composite materials using the NASA CFRP dataset.
(2) We have proposed a deep transfer learning scheme to predict damage for a test coupon based on a model pretrained on training coupons.
(3) The proposed method leverages existing fatigue data from different composite coupons by training a TCN on the combined dataset to generate a pretrained TCN for composite damage prediction.
(4) We have shown that the proposed transfer learning scheme (TCN-trans) enables damage prediction for a new composite coupon by utilizing fewer samples of the test coupon and by training for much shorter time.

The paper is organized as follows. Section 2 describes the TCN model and the transfer learning scheme. Section 3 presents the results of TCN parameter selection and the application of transfer learning (TCN-trans) on NASA carbon fiber reinforced polymer (CFRP) sensor data for composite failure [2] and Sect. 4 provides the conclusions from this study.

2   Methodology

In this section, TCN-trans framework used for damage prediction is described in detail. A TCN is based on two principles: 1) there is no "leakage" of information from future into the past, and 2) an input sequence of any length can be mapped into an output sequence of the same length by the TCN architecture. Next, we describe how different components of a TCN incorporate these principles.

2.1   Causal Dilated Convolution

To prevent the leakage of future information into the past, TCN employs causal convolutions. It involved convolving an output at time t with elements at time t and earlier in the previous layer. Further, in order to produce output of the same length as input, and to achieve a long effective memory size, TCN employs dilated convolutions. Dilated convolutions involve increasing the size of convolution kernel by a predefined dilation factor, while keeping the number of weights unchanged, and therefore support the expansion of receptive field of the network and increasing its representation ability [14]. Hence, d-dilated convolution (F) on element s of a 1-D input sequence $\mathbf{x}$, for filter f, is defined as:

$$F(s) = (x *_d f)(s) = \sum_{i=0}^{k-1} f(i) \cdot x_{s-d \cdot i} \qquad (1)$$

where d is the dilation factor, and k is the filter size. A dilation factor of 1 would imply a regular convolution. When d-dilated convolutions are stacked in an exponential manner, such that $d = 2^n$ for the nth layer, a long effective memory can be achieved [15]. Figure 1 (left) illustrates causal dilated convolution where kernel size used is $k = 3$ and dilation factor, $d = 1, 2$. The dilation factor used is systematically increased by a factor of 2 until the receptive field exceeds the input sequence length.

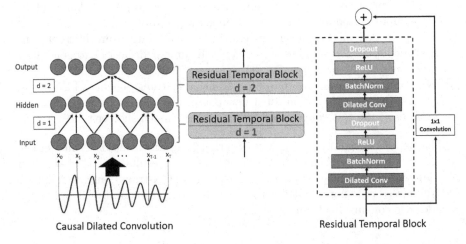

Fig. 1. Left panel shows dilated convolution where kernel size used is k = 3 and dilation factor of d = 1, 2. Dilated convolutional layers incorporated into a residual temporal block (right), with residual blocks stacked up to make a TCN (middle).

2.2   Residual Temporal Block

The convolutional layers of TCN are stacked into residual temporal blocks (Fig. 1 (middle), where the outputs of a series of transformations G are added to the input $\mathbf{x}$ of the residual block [16], as represented by the equation below:

$$o = Activation(\mathbf{x} + G(\mathbf{x})) \qquad (2)$$

Thus, the output of a residual block is computed relatively with respect to an input. Each residual block of a TCN consists of two convolutional layers of identical dilation factor, each followed by weight normalization, rectified linear unit (ReLU) activation and spatial dropout for regularization (Fig. 1(right)). A residual connection which adds the input of the residual block to the output of the final activation layer is also introduced. A 1 × 1 convolution layer is added in the residual connection to ensure the dimensions of the input and output tensors are compatible for addition. Multiple residual blocks are stacked in order of increasing dilation factor and an addition operation is then performed on the output of each residual block, forming the input to the recurrent component.

2.3   TCN Based Transfer Learning (TCN-trans):

We employ the TCN for transfer learning by using a pretrained model for predictions on a new dataset. Here, TCN is first trained on a source dataset for N epochs and the weights of the trained model are saved. Next, TCN is initialized with the saved weights and the model is fine-tuned by training on the target dataset for smaller number of epochs (N_1) using less number of samples. This scheme is explained in Fig. 2.

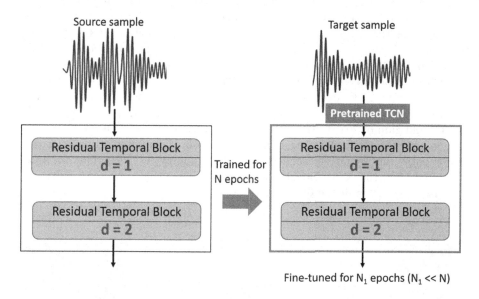

Fig. 2. TCN based transfer learning scheme: TCN-trans.

3   Results

In this section, the experimental results are presented. First, the CFRP dataset and the preprocessing techniques are described. Next, the performance of the TCN model is evaluated on the CFRP dataset. We study the effect of model parameters and input parameters on the predictive performance of TCN and finally, apply the transfer learning method for predicting delamination of a composite coupon.

3.1   Data

In this paper, TCN model is applied to the sensor data for composite failure from the NASA Prognostics Data Repository [2]. This dataset comprises run-to-failure measurements from Lamb wave sensors and X-ray images of specimens for capturing the propagation of damage in carbon fiber composite coupons under tension-tension fatigue loading. The sensor data was collected from six composite coupons made of Torayca T700G uni-directional carbon prep-reg material in dog-bone geometry and a notch to induce stress concentration. Three symmetric layup configurations were used for these coupons to account for the effects of ply orientation: Layup 1: $[0_2/90_4]$ for coupon denoted by L1S19, Layup 2: $[0/90_2/45/-45/90]$ for coupons L2S17, L2S18, L2S20, and Layup 3: $[90_2/45/-45]_2$ for coupons L3S18, L3S20.

As shown in Fig. 3(a), two sets of six PZT sensors, with six actuators (denoted by numbers 1 to 6) and six sensors (denoted by numbers 7 to 12), attached on both ends of the coupons were used to monitor propagation of Lamb waves through the samples. A sampling rate of 1.2 MHz was used to acquire signals from all of the 36 actuators-sensors for a duration of 1667 microseconds, resulting in 2000 data points for each measured signal (Fig. 3(b)). Lamb wave sensing was done using a five cycle tone-burst actuation, which was excited at seven interrogation frequencies (F) in the range of 150–450 kHz at an average amplitude of 50 V and gain of 20 dB. Finally, different loading cycles were used during the fatigue test, thereby generating Lamb wave sensor data for multiple actuator-sensor paths, interrogation frequencies and loading cycles. Delamination area, estimated from the X-ray images (Fig. 3(c)) using image processing software, was also provided with the CFRP dataset.

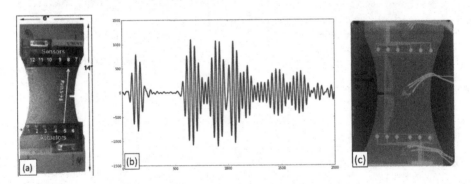

Fig. 3. (a) Coupon specimen, actuator-sensor path 5–8 shown with the white arrow. (b) Lamb wave signal recorded at a sensor on the coupon. (c) X-ray image of coupon L1S11 (not used in this work) after 100 K cycles of fatigue loading. The delamination damage is shown as light gray area centered around the notch.

3.2 Preprocessing and Training

The sensor data recorded from 6 different coupons was combined and then transformed by performing standard scaling, by subtracting the mean and dividing by the standard deviation. The TCN was trained on sensor data by dividing the dataset into training data comprising 80% of the data samples and validation data consisting of the remaining 20% samples. The prediction results are reported for the validation dataset. The model was trained using RMSprop optimizer, which uses a momentum term to enable faster convergence of the algorithm, and another parameter to prevent the gradients from blowing up. TCN was trained for 500 epochs using a batch size of 16, dropout rate of 0.2 and learning rate of 0.005. Our experimental platform is a server with NVIDIA GeForce RTX 2080 Ti GPU with Ubuntu 18.04 operating system. The programming language used in this work is Python 3.7 with deep learning library PyTorch (1.1.0).

3.3 TCN Parameter Selection

We first study the effect of varying the parameters of TCN model on the prediction of delamination areas, A. For this study, 3 parameters – hidden size (h), kernel size (k) and dilation factor (d) were varied. We report Root Mean Square Error (RMSE) and R-squared or coefficient of determination (R^2) values for various combinations of the model parameters and F $= 300$ kHz, in Table 1. We can see that the model performs better (low RMSE, high R^2) as dilation factor increases, with the smallest RMSE and largest R^2 values obtained for h $= 100$, k $= 7$ and d $= 8$. However, model performance and choice of model parameters can vary with changing input parameters like, signal sequence length (L) and interrogation frequency (F). Next, we investigate the effects of these parameters on the prediction of delamination damage.

Table 1. TCN parameter selection results showing RMSE and R^2 values corresponding to delamination area prediction

Kernel size	Dilation factor	h $= 30$	h $= 50$	h $= 100$
k $= 5$	d $= 2$	636.86, 0.31	636.48, 0.31	635.80, 0.31
k $= 5$	d $= 4$	632.65, 0.32	631.03, 0.32	631.30, 0.32
k $= 5$	d $= 8$	383.25, 0.75	383.42, 0.75	369.28, 0.80
k $= 7$	d $= 2$	637.22, 0.31	636.08, 0.31	639.84, 0.30
k $= 7$	d $= 4$	624.36, 0.34	631.24, 0.32	627.81, 0.33
k $= 7$	d $= 8$	363.05, 0.78	329.55, 0.82	**325.86, 0.84**
k $= 12$	d $= 2$	634.03, 0.32	627.95, 0.33	627.10, 0.33
k $= 12$	d $= 4$	616.67, 0.35	615.33, 0.36	622.42, 0.34
k $= 12$	d $= 8$	367.07, 0.77	334.76, 0.81	365.19, 0.78

3.4 Effect of Input Parameters on Delamination Prediction

In order to determine which part of the Lamb wave signal is most sensitive to delamination damage, different lengths of the sensor signal (L) were used –

200, 500, 1000, 1500 and 2000, where the sensor signal consists of a total of 2000 data points. Here, sequence length corresponds to the first L points of the Lamb wave signal. Moreover, we also studied the effect of frequency at which Lamb wave signals were sensed by varying F from 150, 200, 300, 350 to 450 kHz. The TCN model was trained for different combinations of L and F by selecting the model parameters (h, k, d) giving the best results. Figure 4 showing the heatmap for the best R^2 results for each pair of L, F values indicates that Lamb wave signal based delamination prediction improves as interrogation frequency increases with higher R^2 (> 0.8) attained for F $\geq$ 300 kHz. Moreover, these predictions also depend on the sequence length with L $\geq$ 800 data points of the Lamb wave important for F = 300 kHz, L $\geq$ 1000 for F = 350 kHz and L $\geq$ 1500 for F = 450 kHz, as shown within the green boundary.

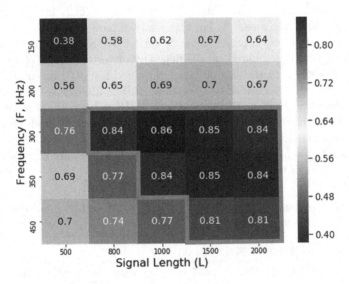

Fig. 4. Heatmap showing R^2 scores for delamination area prediction for different sequence lengths (L) and frequencies (F).

In order to further understand, which L and F values can be most relevant for delamination area prediction for different coupons, we looked at the coupon specific performance. We plotted the RMSE values for different coupons at three best frequencies (F = 300, 350, 450 kHz) corresponding to L = 1500, in Fig. 5(a). We found that TCN based delamination prediction for all coupons was similar at both 300 and 350 kHz, while the corresponding RMSE values were higher at 450 kHz. However, the R^2 scores plotted in Fig. 5(b) show that the model does not perform well for each coupon, particularly for L2S20 and L3S18 coupons, which have low or even negative R^2 scores. In order to address this issue, we propose a transfer learning scheme based on fine-tuning a pretrained TCN on new coupon data. This is explained and corresponding results are discussed in the next section.

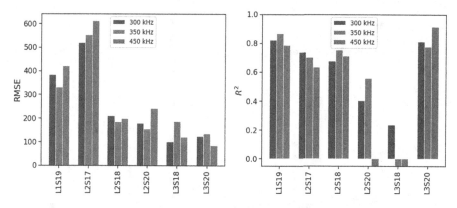

Fig. 5. Bar plots showing RMSE (left) and R^2 scores (right) for different coupons at $L = 1500$ and $F = 300, 350, 450\,$kHz.

3.5   TCN-trans: TCN Based Transfer Learning for Delamination Prediction

The transfer learning scheme involves using a pretrained TCN for predicting the delamination damage of a coupon. For this study, we trained TCN on sensor data from 5 of the 6 coupons for 500 epochs and treated this as a "pretrained" model. For the 6th test coupon, delamination area, A, was predicted using the Lamb wave signal as input to the model. We tested and compared the predictions of 3 schemes for this task: 1) Direct transfer - the pretrained model was directly used to predict delamination for the test coupon; 2) Fine-tuned - 80% of samples from the test coupon data were used to fine-tune the pretrained model by training it for 100 epochs; and 3) Retrain - 80% of the test coupon data was used to train TCN from scratch, for 500 epochs. This process was repeated for each coupon and the corresponding results are shown in Fig. 6. The RMSE values of delamination area predicted for each coupon show that the fine-tuned TCN model performs much better than the direct transfer scheme, where the pretrained model has not seen the new test coupon data. Moreover, the fine-tuned TCN yields better predictions than a TCN retrained only on the test data.

We also looked at the number of samples of test data required for fine-tuning the pretrained TCN and its effect on the model performance. For this, we used 30%, 50%, 60% and 80% (as in Fig. 6, orange bar) of the test coupon data for fine-tuning the pretrained TCN. Figure 7 shows the RMSE and R^2 values for L1S19 test coupon using TCN pretrained on remaining 5 coupon data and further fine-tuned using different numbers of test samples. As more number of test samples are used for fine-tuning the model, RMSE reduces and R^2 increases. Moreover, we can see that the pretrained TCN can achieve $R^2 \geq 0.8$ by utilizing only 50–60% of the test samples for fine-tuning the model.

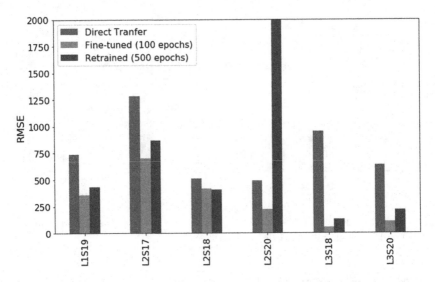

Fig. 6. RMSE for each coupon corresponding to three schemes - direct transfer from pretrained TCN, fine-tuned TCN on test data, TCN retrained on only test data.

4 Conclusions

We have proposed a temporal convolutional network based transfer learning (TCN-trans) method for composite damage prediction. This proposed framework was applied on CFRP dataset consisting of Lamb wave signals from a network of 36 PZT sensors for different composite coupons subjected to cyclic fatigue loading. Firstly, we trained the TCN on data collected from six different coupons having three different layup configurations and showed the effect of input parameters like interrogation frequency and length of the Lamb wave signal on delamination size prediction. We showed that the model trained on data from different coupons may not perform well for each coupon, pointing to the importance of transfer learning in such a scenario. Hence, we applied the TCN based transfer learning approach in which a TCN pretrained on source dataset consisting of Lamb wave signals from five coupons with different layup configurations, was used to predict delamination area for a target/test coupon.

We showed that a pretrained TCN, when fine-tuned using test coupon data for much less number of training epochs, can achieve better predictions than a TCN trained only on the test data for much longer training time. Further, we also demonstrated that reasonably good predictions can be attained with fine-tuning the pretrained TCN with only 50–60% of the test samples for much less number of epochs. This kind of scheme becomes extremely important for deep learning applications in aerospace or manufacturing domains, which suffer from the lack of sufficient labeled experimental data. Moreover, transfer learning based structural damage prediction is highly relevant in the field of composite failure analysis, where transfer learning can alleviate the need of training a model

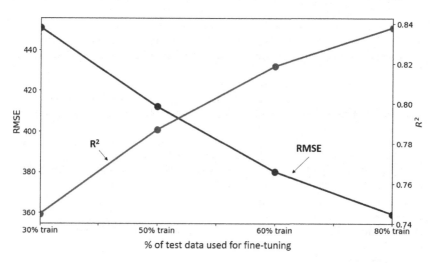

Fig. 7. RMSE and R^2 scores obtained for L1S19 coupon using the fine-tuned TCN as a function of test data size used for fine-tuning.

fully from scratch every time the fatigue performance of a different specimen is being studied.

References

1. Johnson, P., Chang, F.-K.: Characterization of matrix crack-induced laminate failure part I: experiments. J. Compos. Mater. **35**(22), 2009–2035 (2001)
2. Saxena, A., Goebel, K., Larrosa, C.C.: Accelerated aging experiments for prognostics of damage growth in composite materials. In: Proceedings of the 8th International Workshop on Structural Health Monitoring, Stanford, CA, USA, vol. 1, pp. 1139–1149 (2011)
3. Ihn, J.-B., Chang, F.-K.: Pitch-catch active sensing methods in structural health monitoring for aircraft structures. Struct. Health Monit. **7**(1), 5–19 (2008)
4. Kessler, S.S., Spearing, S.M., Soutis, C.: Damage detection in composite materials using Lamb wave methods. Smart Mater. Struct. **11**(2), 269–278 (2002)
5. Lamb, H.: On waves in an elastic plate. Proc. Roy. Soc. London, Ser. A **93**D, 114–128 (1917)
6. Su, Z., Ye, L., Lu, Y.: Guided lamb waves for identification of damage in composite structures: a review. J. Sound Vib. **295**(3), 753–780 (2006)
7. Mitra, M., Gopalakrishnan, S.: Guided wave based structural health monitoring: a review. Smart Mater. Struct. **25**(5), 053001 (2016)
8. Liu, H., Liu, S., Liu, Z., Mrad, N., Dong, H.: Prognostics of damage growth in composite materials using machine learning techniques. In: IEEE International Conference on Industrial Technology (ICIT), pp. 1042–1047 (2017)
9. Mishra, S., Vanli, O.A.: Remaining useful life estimation with lamb-wave sensors based on Wiener process and principal components regression. J. Nondestructive Eval. **35**(1), 1–13 (2016). https://doi.org/10.1007/s10921-015-0328-2

10. Lahmadi, A., Terrissa, L., Zerhouni, N.: A data-driven method for estimating the remaining useful life of a composite drill pipe. In: International Conference on Advanced Systems and Electric Technologies (IC_ASET), Hammamet, pp. 192–195 (2018)
11. LeCun, Y., Bengio, Y., Hinton, G.: Deep learning. Nature **521**, 436–444 (2015)
12. Tan, C., Sun, F., Kong, T., Zhang, W., Yang, C., Liu, C.: A survey on deep transfer learning. In: Proceedings of the 27th International Conference on Artificial Neural Networks, Rhodes, Greece (2018)
13. Ansi, Z., et al.: Transfer learning with deep recurrent neural networks for remaining useful life estimation. Appl. Sci. **8**(12), 2416 (2018)
14. Oord, A.V.D., et al.: WaveNet: a generative model for raw audio. Arxiv (2016)
15. Bai, S., Kolter, J.Z., Koltun, V.: An empirical evaluation of generic convolutional and recurrent networks for sequence modeling. arXiv preprint arXiv:1803.01271 (2018)
16. He, K., Zhang, X., Ren, S., Sun, J.: Deep residual learning for image recognition. In: 2016 IEEE Conference on Computer Vision and Pattern Recognition (CVPR), Las Vegas, NV, pp. 770–778 (2016)

Dilated Convolutional Recurrent Deep Network with Transfer Learning for Remaining Useful Life Prediction

Jing Yang Lee[1], Ankit K. Das[2], Shaista Hussain[2(✉)], and Yang Feng[2]

[1] Nanyang Technological University, Singapore, Singapore
[2] Institute of High Performance Computing, ASTAR, Singapore, Singapore
hussains@ihpc.a-star.edu.sg

Abstract. Remaining Useful Life (RUL) prediction of industrial systems/components helps to reduce the risk of system failure as well as facilitates efficient and flexible maintenance strategies. However, due to cost and/or time limitations, it is a major challenge to collect sufficient lifetime data, especially the fault/failure data, from an industrial system for training an efficient model to predict the RUL. In addition, many of these systems work under dynamically changing operating conditions, which would require model retraining for adapting and generalizing to new conditions. In order to address these issues, we propose an architecture comprising a Dilated Convolutional Neural Network, which utilises non-causal dilations, combined with a Long Short-Term Memory Network: DCNN-LSTM model for RUL prediction. This model was validated on the publicly available NASA turbofan dataset and its performance was benchmarked against previously proposed models, showing the improvement by our proposed model. Next, DCNN-LSTM model was used in a transfer learning setting where both the issues of model retraining and limited availability of experimental data were addressed. The results showed a significant reduction in training time compared to the time required for retraining models from scratch, while achieving similar performance. Moreover, the proposed method also achieved at par performance by utilizing a much smaller amount of data.

Keywords: Remaining Useful Life · Dilated convolutional recurrent neural networks · Prognostics

1 Introduction

Prognostics refers to a field of engineering that involves predicting the Remaining Useful Life (RUL) of systems under various fault and operational conditions [1]. Prognostic methods can be broadly classified as either data-driven or physics based [2]. Physics based methods leverage on mathematical modelling in order to estimate the RUL. On the other hand, in data-driven prognostics, machine learning algorithms are used to model the multivariate time series data generated

© Springer Nature Switzerland AG 2020
W. Lu and K. Q. Zhu (Eds.): PAKDD 2020 Workshops, LNAI 12237, pp. 153–164, 2020.
https://doi.org/10.1007/978-3-030-60470-7_15

by numerous sensors. Data-driven approaches have achieved considerable success and widespread recognition in recent years [3]. This is due to the success of machine learning algorithms and the increased availability of industrial sensory data [3]. However, many traditional machine learning algorithms rely on manual feature extraction for data representation. Another class of machine learning algorithms known as deep learning algorithms automatically extracts features from raw data without the need for expert knowledge.

In recent years, many deep learning algorithms have been proposed for RUL prediction [4–7]. In [4], a Convolutional Neural Network (CNN) featuring two convolutional layers followed by max pooling layers has been proposed. However, the CNN based algorithm does not take into consideration the sequential element of the multivariate time series data. Recurrent Neural Network (RNN) are a class of algorithms that predict the output as a function of current input and previous states thereby preserving the sequential information. Since normal RNNs suffer from the exploding and vanishing gradient problem [8], LSTMs and GRUs have become synonymous with multivariate time series prediction tasks [5]. Hsu et al.[6] demonstrated the effectiveness of LSTMs by in predicting RUL using the CMAPSS dataset with a model that consists of only vanilla LSTMs. Wang et al. [7] implemented a model incorporating two BiLSTM layers for RUL prediction allowing the model to utilize the sequence in both directions. Jayasinghe et al. [9], on the other hand, developed a hybrid model incorporating both CNN and LSTM for RUL prediction on the C-MAPSS dataset. This model consists of three pairs of convolutional and pooling layers, two LSTM layers and three fully connected layers. However, the lack of sufficient experimental data limits the usage of deep learning algorithms in this field. Transfer learning algorithms are a class of algorithms that transfer latent knowledge learnt from a source to a target domain [10]. Zhang et al. [11] implemented bidirectional LSTM models for transfer learning by first training on a source data followed by using fewer samples from target data set. Moreover, another challenge is that these algorithms need to be retrained in case of systems/components working under dynamic operating conditions resulting in large training time.

In order to address these issues, we propose a Dilated CNN-LSTM (DCNN-LSTM) with transfer learning for RUL prediction using the NASA C-MAPSS dataset [12]. The dilations in the proposed network supports exponential expansion of the receptive field thereby increasing its representation ability [13]. While causal dilations have been a key component in temporal CNNs which have been largely used as an alternative to recurrent models [13,14], our proposed model supplements the LSTM component by performing feature engineering using non-causal dilations. The LSTM component accounts for the sequential information in the data. In particular, the DCNN-LSTM model comprises of a dilated convolution component followed by a recurrent component. The dilated convolution component contains dilated convolutional layers incorporated into numerous residual blocks. The recurrent component constitutes a pair of stacked LSTMs. Next, we employ this architecture for transfer learning in two different settings. First, we address the issue of model retraining by fine-tuning a pretrained model on a source dataset as opposed to training a new model from scratch. This will alleviate the need to retrain the model every time there is a new dataset under existing or new operating condition. Secondly, the issue of limited experimental data is addressed by fine-tuning a pretrained model on a much smaller dataset, thereby mitigating the requirement of large amount training data. Our results show that a significant reduction in training time can be achieved as compared to the time required for retraining models from scratch, while achieving similar performance. Moreover, we also showed that DCNN-LSTM achieves at par performance by utilizing much less amount of data.

The remainder of this paper is organized as follows: Sect. 2 describes the implemented DCNN-LSTM model. Section 3 presents the results of DCNN-LSTM model against previously implemented models and evaluates the results of transfer learning. Section 4 provides the conclusions from this study.

2 Methodology

In this section, the proposed DCNN–LSTM model will be described in detail. Figure 1 shows the DCNN-LSTM model. It comprises two main components: dilated convolution component and the recurrent component. The dilated convolution component contains dilated convolutional layers incorporated into numerous residual blocks. The recurrent component constitutes a pair of stacked LSTMs. Next, we provide details of each component in the subsequent sections.

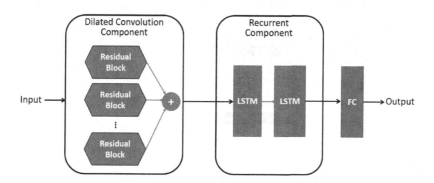

Fig. 1. Overview of DCNN-LSTM model

2.1 Dilated Component

Dilated Convolutions: Dilated convolutions involve increasing the size of convolution kernel by a predefined dilation factor, while keeping the number of weights unchanged. A dilation factor of 1 would imply a regular convolution. There are two forms of dilations: causal and non-causal. Causal dilation restricts the convolution filter at a particular time step T from including any input later than T in the convolution operation. On the other hand, non-causal dilation filters include time steps before and after T in the convolution operation. In our DCNN-LSTM model, feature engineering is performed via non-causal dilation. The formula for non-causal dilation on an arbitrary 1D sequence (s) is defined in Eq. 1. df refers to the dilation factor, n refers to the filter size, and $s \pm d\Delta i$ captures the non-causal aspect of the dilation operation. Figure 2(a) illustrates a non-causal dilated convolution with kernel size of 3 and dilation factors of $1, 2$. The dilation factor used is systematically increased by a factor of 2 until the receptive field exceeds the input sequence length. In this work, we have used kernel size = 2 and dilation factors = $1, 2, \cdots, 32$ yielding receptive field size of 64. We hypothesize that the non-causal dilation would enhance each time step in the time series with additional relational and temporal information, which would increase the accuracy of the predicted RUL.

$$F(s) = (x *_{df} f)(s) = \sum_{i=0}^{n-1} f(i) \cdot x_{s \pm df \cdot i} \tag{1}$$

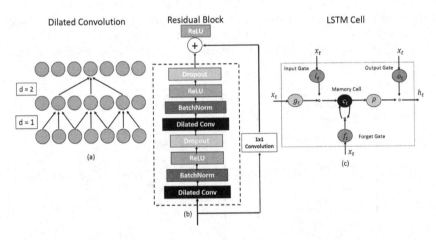

Fig. 2. (a) Dilated convolution with kernel size = 3 and dilation factor = 1, 2. (b) Dilated convolution layers incorporated into a residual block. (c) Overview of an LSTM cell.

Residual Block: In the dilated component, dilated convolutions are incorporated into residual blocks. Each residual block consists of two convolution layers of identical dilation factor, each followed by batch normalization, ReLU activation and spatial dropout layers Fig. 2(b). A spatial dropout rate of 0.5 is used in our model. A residual connection which adds the input of the residual block to the output of the final activation layer is also introduced. A 1×1 convolution layer is added in the residual connection to ensure the dimensions of the input and output tensors are compatible for addition. A final ReLU activation layer is also introduced after the addition operation. Multiple residual blocks are then stacked in order of increasing dilation factor. The dilation factors used in the DCNN-LSTM model are $1, 2, \cdots, 32$. An addition operation would then be performed on the output of each residual block, forming the input to the recurrent component.

2.2 Recurrent Component

Long Short Term Memory (LSTM) Network: LSTMs are a class of RNNs that are designed to solve the vanishing and exploding gradient problems that are associated with regular RNNs. LSTM achieves this by introducing a forget gate in conjunction with the input and output gates. Figure 2(c) presents an overview of the operations within an LSTM cell. The input gate i_t, output gate o_t, forget gate f_t constitute the main components of an LSTM node. The memory cell and the output are denoted by c_t and h_t, respectively. The operations of the LSTM cell are described by the Eqs. 2. The input, output, forget gate and cell state weight matrices and input, output, forget gate and cell state biases are denoted by W_i, W_o, W_f, W_c and b_i, b_o, b_f, b_c, respectively. These parameters are shared across time steps and tuned throughout the training process. σ refers to the sigmoid activation function and $\cdot$ refers to the element wise multiplication operator. In our recurrent component, 2 LSTM cells of 56 hidden units are used. A dropout rate of 0.25 is used in the first LSTM cell, and a dropout rate of 0.35 is used in the second LSTM cell.

$$g_t = \tanh(W_c X_t + U_c h_t + b_c)$$
$$o_t = \sigma(W_o X_t + U_o h_{(t-1)} + b_o)$$
$$i_t = \sigma(W_i X_t + U_i h_{(t-1)} + b_i)$$
$$f_t = \sigma(W_f X_t + U_f h_{(t-1)} + b_f)$$
$$c_t = f_t c_{(t-1)} + i_t g_t$$
$$h_t = o_t + \tanh(c_t) \tag{2}$$

2.3 DCNN-LSTM with Transfer Learning

We employ the proposed DCNN-LSTM model for transfer learning. Figure 3 shows the transfer learning scheme employed in this work. Firstly, the target model would be pretrained on the source dataset for a fixed number of epochs ($epochs_{pretrain}$). Then the model would be fine tuned on the target dataset. This allows the target model to be effectively initialized with the weights learned during training on the source dataset. This framework is formalized in Algorithm 1.

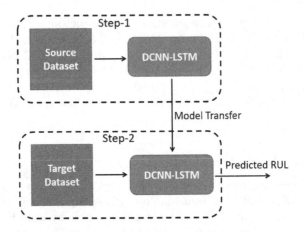

Fig. 3. Framework for transfer learning

Algorithm 1. Transfer Learning Framework

1: Preprocess both Source and Target dataset
2: **for** $epochs = 1, 2, 3, \ldots, epochs_{pretrain}$ **do**
3:     Pretrain DCNN-LSTM model on Source dataset
4: **end for**
5: Fine tune DCNN-LSTM model on Target Dataset

---

3   Experiment Results

In this section, the experimental results are presented. First, the C-MAPSS dataset and the preprocessing techniques are described. Next, the performance of the proposed DCNN-LSTM model is evaluated on the C-MAPSS dataset. The performance of DCNN-LSTM is compared with other algorithms including Multilayer Perceptron (MLP) [4], Support Vector Regression (SVR) [4], Relevance Vector Regression (RVR) [4], CNN [4] and LSTM [6]. Finally, the performance of DCNN-LSTM is evaluated in the two transfer learning settings.

3.1   DCNN-LSTM Performance Evaluation

Data: In this paper, the NASA Commercial Modular Aero-Propulsion System Simulation (C-MAPSS) dataset [12] is used. The C-MAPSS dataset consists of four sub datasets. As seen in Table 1, each sub dataset contains multivariate time series trajectories of either 1 or 6 operational conditions and 1 or 2 fault conditions. Every sub dataset is also split into a training and test dataset. Each multivariate time series consists of the id, cycle number, three operational settings and 21 sensor readings. The training datasets comprise trajectories where the fault grows with each cycle, culminating in system failure. On the other hand, the test dataset contains trajectories up to some time prior to system failure.

Table 1. Sub datasets of the C-MAPSS dataset

	FD001	FD002	FD003	FD004
Operating conditions	1	6	1	6
Fault conditions	1	1	2	2
Trajectories (Train)	100	260	100	249
Trajectories (Test)	100	259	100	248

Pre-processing: A sliding window approach is adopted in this work in order
to produce sequential data for training. A window size of 50 is used to sample
data from the training dataset. The time window is then shifted by one cycle
to generate a new sample and this process is repeated up to the end-of-life of
the engine to yield multiple multivariate time series with sequence length of 50.
The sensor data was normalized by transforming each feature in the range of
0–1 (refer Eq. 3).

$$x^{'} = \frac{x - min(x)}{max(x) - min(x)} \tag{3}$$

Model Hyperparameters: The hyperparameters used in the dilated convolution component and recurrent component of the DCNN-LSTM model are summarised in Table 2. The number of filters, kernel size, and dropout rate used in
the dilated convolution component are kept constant over all convolution layers
in the residual blocks. Mean squared error was used as the loss function during
training of the DCNN-LSTM model.

Table 2. Breakdown of hyperparameters used in DCNN-LSTM model.

Dilated convolution component				
Num of residual blocks	Dilation factors	Num of filters	Kernel size	Dropout
6	1, 2, 4, 8, 16, 32	28	2	0.5
Recurrent component				
Hidden units (LSTM1)	Dropout (LSTM1)	Hidden units (LSTM2)	Dropout (LSTM2)	
56	0.25	56	0.35	

Performance Evaluation: The performance of the DCNN-LSTM model is
evaluated and compared with existing algorithms in the literature. We employ
the Root Mean Squared Error (RMSE) to compare the performance of all the
algorithms. Table 5 shows the RMSE of the evaluated algorithms. It can be seen
from the table that DCNN-LSTM performs better than other algorithms on
datasets on FD001, FD002 and FD003. In case of FD004 the performance is
comparable with LSTM and CNN (Table 3).

Table 3. Performance comparison (RMSE) of DCNN-LSTM with MLP, SVR, RVR, CNN, LSTM.

Sub Dataset	FD001	F002	FD003	FD004
MLP [4]	37.56	80.03	37.39	77.37
SVR [4]	20.96	41.99	21.05	45.35
RVR [4]	23.80	31.30	22.37	34.34
CNN [4]	18.45	30.29	19.82	29.16
LSTM [6]	16.73	29.43	18.07	**28.39**
DCNN-LSTM	**14.08**	**23.44**	**16.08**	31.15

3.2 DCNN-LSTM with Transfer Learning

Next, we evaluate the performance of the DCNN-LSTM model in transfer learning settings. First, we employ the pretrained DCNN-LSTM on a target dataset to avoid training a new model from scratch. Next, we evaluate the pretrained DCNN-LSTM model on limited number of trajectories of the target dataset. A list of the variables used in this section and their corresponding descriptions are described in Table 4.

Table 4. Description of variables used in Sect. 3.2

Variable	Description
$epochs_{pretrain}$	Number of epochs used to pretrain DCNN-LSTM on source dataset
$epochs_{trans}$	Number of epochs required to finetune DCNN-LSTM transfer model on target dataset
$RMSE_{trans}$	RMSE achieved by DCNN-LSTM transfer model on target dataset after finetuning
$epochs_{notrans}$	Number of epochs required to train DCNN-LSTM directly on target dataset without transfer
$RMSE_{notrans}$	RMSE achieved by DCNN-LSTM on target dataset without transfer
IMP_{epochs}	Percentage improvement (Eq. 4) in $epochs_{trans}$ over $epochs_{notrans}$.

$$IMP_{epochs} = \frac{epochs_{notrans} - epochs_{trans}}{epochs_{notrans}} \times 100 \qquad (4)$$

DCNN-LSTM with Transfer Learning for Reduction in Training Time: We evaluated all 12 transfer pairs among the 4 sub datasets (e.g. FD001 → FD002, FD003 → FD004) in the C-MAPSS dataset. The right arrow (→) denotes the source and target for transfer learning. For example, in the case of FD001 → FD002, the DCNN-LSTM model will be pretrained on FD001 (source) and fine tuned on FD002 (target). Table 5 shows the $epochs_{notrans}$, $RMSE_{trans}$, $epochs_{notrans}$, $RMSE_{notrans}$, $epochs_{pretrain}$ and IMP_{epochs} of the respective transfer pairs. In this experiment, the model has been pretrained on the source dataset ($epochs_{pretrain}$) for 10, 20, 30 and 40 epochs. It can be seen from the table that the transfer learning scheme resulted in a significant decrease in the number of epochs or $epochs_{trans}$, implying a reduction in training time. This is most prominently demonstrated in both FD001 → FD003 and FD003 → FD001 where $epochs_{trans}$ improved by 84.31% and 82.98%, respectively, after fine-tuning a model that was pretrained on the source dataset for 40 epochs. It should be noted that $epochs_{pretrain}$ and IMP_{epochs} are directly proportional, generally. The higher the number of pretraining epochs, the greater the reduction in number of training epochs, resulting in a larger IMP_{epochs} percentage value. Figure 4 shows the comparison between RMSE values for models pretrained on different datasets. It can be observed that pre-training the model on a source dataset and fine tuning on a target dataset results in similar RMSE values as training the model from scratch on the target dataset. It is also observed that transfer learning from simpler datasets to more complex datasets (identical operational conditions or fault conditions) generally results in better performing transfer models with lower $RMSE_{trans}$ values.

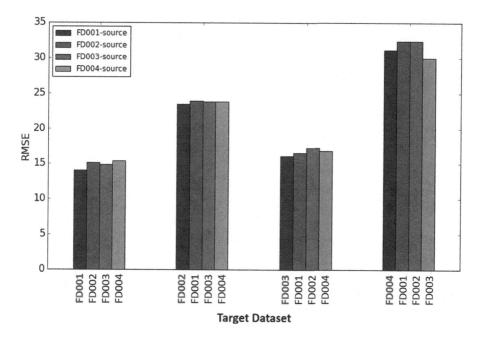

Fig. 4. Bar plots depicting $RMSE_{notrans}$ (blue bars) and the $RMSE_{trans}$ of each transfer pair. (Color figure online)

Table 5. Transfer learning results showing improvement in training time using pre-trained DCNN-LSTM model.

Transfer pair	Performance before transfer		Performance after transfer			
	$epochs_{notrans}$	$RMSE_{notrans}$	$epochs_{pretrain}$	$epochs_{trans}$	IMP_{trans}	$RMSE_{trans}$
FD001 → FD002	181	23.44	10	203	−12.15	23.64
			20	127	29.83	23.96
			30	**44**	**75.69**	**23.95**
			40	40	77.90	24.96
FD001 → FD003	51	16.08	10	40	21.57	16.66
			20	**28**	**45.10**	**16.52**
			30	19	62.75	16.91
			40	8	84.31	17.99
FD001 → FD004	157	31.15	**10**	**123**	**21.65**	**34.64**
			20	87	44.59	33.97
			30	52	66.88	35.09
			40	43	72.61	32.46
FD002 → FD001	47	14.08	**10**	**32**	**31.91**	**15.15**
			20	36	23.40	16.26
			30	25	46.81	16.10
			40	24	48.94	16.73
FD002 → FD003	51	16.08	10	47	7.84	17.06
			20	39	23.53	17.87
			30	**33**	**35.29**	**17.25**
			40	42	17.64	18.35
FD002 → FD004	157	31.15	10	163	−3.82	32.75
			20	115	26.75	34.23
			30	101	35.67	33.58
			40	**94**	**40.13**	**32.46**
FD003 → FD001	47	14.08	**10**	**28**	**40.43**	**14.88**
			20	21	55.32	15.20
			30	15	68.09	14.95
			40	8	82.98	15.11
FD003 → FD002	181	23.44	10	197	−8.84	23.84
			20	**129**	**28.73**	**23.86**
			30	112	38.12	25.49
			40	132	27.07	24.09
FD003 → FD004	157	31.15	10	121	22.93	32.12
			20	65	58.60	29.91
			30	**46**	**70.70**	**29.96**
			40	30	80.89	30.83
FD004 → FD001	47	14.08	10	31	34.04	15.75
			20	**24**	**48.94**	**15.41**
			30	22	53.19	17.19
			40	20	57.45	18.59
FD004 → FD002	181	23.44	10	199	−9.94	24.23
			20	139	23.20	24.93
			30	126	30.34	24.28
			40	**104**	**42.54**	**23.82**
FD004 → FD003	51	16.08	**10**	**42**	**17.64**	**16.86**
			20	31	39.22	17.02
			30	25	50.98	18.52
			40	20	60.78	19.47

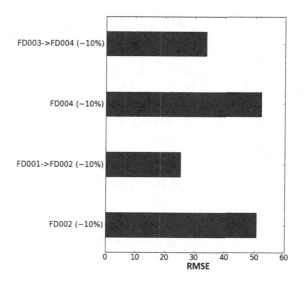

Fig. 5. Transfer learning results showing improved performance (smaller RMSEs) using smaller number of trajectories (10%).

DCNN-LSTM with Transfer Learning for Limited Experimental Data:
In real world RUL prediction tasks, the collection of experimental data is always a challenge. In this section, in order to address this challenge, we evaluated the performance of DCNN-LSTM with transfer learning with limited experiment data. In particular, we evaluate the performance of pretrained models on source dataset by considering smaller fraction (10%) of target dataset trajectories. Here, we consider two better preforming (FD001 → FD002 and FD003 → FD004) transfer pairs from the previous experiment. Figure 5 shows the accuracy of transferred model which is fine-tuned on only 10% of the trajectories, compared with models trained from scratch on only 10% of the trajectories. It can be seen from the figure that in both the cases (FD002 and FD004), DCNN-LSTM with transfer learning achieves much better $RMSE_{trans}$ and performance improvement (~35% and ~50%) compared to cases in which only 10% of data is available for training. The poor performance of models trained from scratch with only 10% of the trajectories could be attributed to overfitting on the training trajectories. This shows that the transfer of knowledge from other datasets using pretrained models could be used in the absence of a sufficiently large dataset.

4   Conclusions

RUL prediction is a key problem in manufacturing domain. This problem is further exacerbated by lack of experimental data and dynamically changing operating conditions which render existing models ineffective. In this paper, we presented a DCNN-LSTM model with transfer learning for RUL prediction of turbofan engines. Comparison with other algorithms such as SVR, MLP, RVR, LSTM and CNN show the efficacy and usefulness of the proposed DCNN-LSTM model. The transfer learning based results showed that reduction in number of training epochs can be achieved by DCNN-LSTM, thereby resulting in shorter training time. Moreover, transfer learning helped achieve similar performance of DCNN-LSTM model even in the absence of large experimental data. In the future, additional research could be dedicated to reducing the number of trainable parameters through the use of the self attention mechanism, which would further decrease training time. The efficacy of transfer learning in self-attention based RUL prediction models should also be explored.

References

1. Kim, N.-H., An, D., Choi, J.-H.: Prognostics and Health Management of Engineering Systems. An Introduction. Springer, Cham (2017). https://doi.org/10.1007/978-3-319-44742-1
2. An, D., Kim, N.H., Choi, J.-H.: Practical options for selecting data-driven or physics-based prognostics algorithms with reviews. Reliab. Eng. Syst. Saf. **133**(C), 223–236 (2015)
3. Zhang, W., Yang, D., Wang, H.: Data-driven methods for predictive maintenance of industrial equipment: a survey. IEEE Syst. J. **13**(3), 2213–2227 (2019)
4. Sateesh Babu, G., Zhao, P., Li, X.-L.: Deep convolutional neural network based regression approach for estimation of remaining useful life. In: Navathe, S.B., Wu, W., Shekhar, S., Du, X., Wang, X.S., Xiong, H. (eds.) DASFAA 2016. LNCS, vol. 9642, pp. 214–228. Springer, Cham (2016). https://doi.org/10.1007/978-3-319-32025-0_14
5. Han, Z., Zhao, J., Leung, H., Ma, K.F., Wang, W.: A review of deep learning models for time series prediction. IEEE Sens. J. **PP**(99), 1 (2019)
6. Hsu, C.-S., Jiang, J.-R.: Remaining useful life estimation using long short-term memory deep learning, pp. 58–61. IEEE (2018)
7. Wang, J., Wen, G., Yang, S., Liu, Y.: Remaining useful life estimation in prognostics using deep bidirectional LSTM neural network, pp. 1037–1042. IEEE (2018)
8. Pascanu, R., Mikolov, T., Bengio, Y.: On the difficulty of training recurrent neural networks. presented at the Proceedings of the 30th International Conference on International Conference on Machine Learning, Atlanta, GA, USA, vol. 28 (2013)
9. Jayasinghe, L., Samarasinghe, T., Yuen, C., Low, J.C.N., Ge, S.S.: Temporal convolutional memory networks for remaining useful life estimation of industrial machinery (2018)
10. Tan, C., Sun, F., Kong, T., Zhang, W., Yang, C., Liu, C.: A survey on deep transfer learning. In: Kůrková, V., Manolopoulos, Y., Hammer, B., Iliadis, L., Maglogiannis, I. (eds.) ICANN 2018. LNCS, vol. 11141, pp. 270–279. Springer, Cham (2018). https://doi.org/10.1007/978-3-030-01424-7_27
11. Ansi, Z., et al.: Transfer learning with deep recurrent neural networks for remaining useful life estimation. Appl. Sci. **8**(12), 2416 (2018)
12. Ramasso, E., Saxena, A.: Performance benchmarking and analysis of prognostic methods for CMAPSS datasets. Int. J. Progn. Health Manag. **5**(2), 1–15 (2014)
13. van den Oord, A., et al.: WaveNet: a generative model for raw audio (2016)
14. Chang, S.-Y., et al.: Temporal modeling using dilated convolution and gating for voice-activity-detection. In: ICASSP (2018)

Generative Modeling for Synthesis of Cellular Imaging Data for Low-Cost Drug Repurposing Application

Shaista Hussain[1(✉)], Ayesha Anees[1], Ankit K. Das[1], Binh P. Nguyen[2],
Mardiana Marzuki[3], Shuping Lin[4], Graham Wright[4], and Amit Singhal[3]

[1] Institute of High Performance Computing, ASTAR, Singapore, Singapore
hussains@ihpc.a-star.edu.sg
[2] School of Mathematics and Statistics, VUW, Wellington, New Zealand
[3] Singapore Immunology Network, ASTAR, Singapore, Singapore
[4] Skin Research Institute of Singapore, ASTAR, Singapore, Singapore

Abstract. Advances in high-content high-throughput fluorescence microscopy have emerged as a powerful tool for several stages of drug discovery process, leading to the identification of a drug candidate with the potential for becoming a marketed drug. This high-content screening (HCS) technology has recently involved the application of machine learning methods for automated analysis of large amount of data generated from screening of large compound libraries to identify drug induced perturbations. However, high costs associated with large-scale HCS drug assays and the limitations of producing abundant high-quality data required to train machine learning models, pose major challenges. In this work, we have developed a computational framework based on deep convolutional generative adversarial network (DCGAN), for the generation of synthetic high-content imaging data to augment the limited real data. The proposed framework was applied on cell-based drug screening image data to derive phenotypic profiles of drug induced effects on the cells and to compute phenotypic similarities between different drugs. Such analysis can provide important insights into repurposing of previously approved drugs for different conditions. Moreover, a generative modeling-based approach of creating augmented datasets can allow to screen more drug compounds within the same imaging assay, thus reducing experimental costs.

Keywords: Deep generative networks · High-content imaging data · Drug repurposing

1 Introduction

High-content screening (HCS) technology enables the combination of high-throughput screening with automated image-based assays and advanced analysis

Supported by ASTAR Joint Council Office (JCO) Career Development Award (CDA) to Amit Singhal [Award number 15302FG151].

W. Lu and K. Q. Zhu (Eds.): PAKDD 2020 Workshops, LNAI 12237, pp. 165–177, 2020.
https://doi.org/10.1007/978-3-030-60470-7_16

tools to facilitate the phenotypic profiling of perturbations induced by thousands of chemical or biological compounds. This approach of screening a large number of drugs/compounds has emerged as a powerful tool in several preclinical stages of drug discovery and development, such as prediction of mechanism of action (MOA), estimation of drug efficacy and toxicity profiling. However, these image-based drug profiling assays used for drug discovery are quite expensive in both academia and pharmaceutical companies. Another issue with such phenotypic screens is that only one assay is screened at a time for a particular disease/condition, which requires multiple replicates of each drug/compound to be screened. These limitations of slow rates and high experimental costs of drug discovery can be addressed by utilizing a repurposing or repositioning approach to transfer the drug-related knowledge gained in one assay to predict biological activity of the drug in another assay. This approach can play a significant role in the drug discovery process by identifying existing drug candidates to be potentially used for new or rare diseases/conditions. Clustering based methods have been used to study the correlations between phenotypic responses and chemical structure of small molecules [16], and such grouping of similar drugs/compounds has been used to predict their off-targets and hypothesize drug-target pairs for potential drug repurposing [2].

In recent years, the use of advanced machine learning algorithms like deep learning models in drug discovery has led to powerful models trained on large amounts of data and has helped to enormously grow the applications of machine learning in pharmaceutical industry. Deep learning algorithms are a class of algorithms that do not rely on hand crafted features and are capable of performing feature extraction from raw data [11], and have achieved state-of-the-art performance in several bioimaging applications, like segmentation of cells and tumor regions, classification of protein subcellular localization and human cell morphological profiles [9,10]. However, deep learning methods require a large number of samples for learning, which becomes a limitation due to the cost associated with data generation, particularly for large-scale drug assays. The challenges in producing abundant high-quality data from preclinical trials can be overcome with the use of a data augmentation approach which involves generating synthetic data samples of high-quality and merging with the limited available data. Synthetic data examples are learned using a generative model, which approximates the distribution of the data space by learning the underlying distribution from the actual samples and use the representation to generate new samples, which are closer to the learned distribution. A class of generative models, Generative Adversarial Networks (GAN) [4], has been used for drug discovery applications, primarily for developing molecular structures [12]. However, the use of generative modeling for augmenting high-content imaging data for drug discovery applications is largely unexplored.

In this work, we present a computational framework for generation of synthetic high-content imaging data, which is used to derive phenotypic similarities between drug-induced perturbations. The proposed framework uses a deep convolutional generative adversarial network (DCGAN) for generating HCS images.

A pre-trained convolutional neural network (CNN) model is used to extract features from both real and synthetic images. The real and synthetic image features are then used to compute drug signatures from the clusters generated using Gaussian mixture model (GMM) method. GMM-based drug signatures are used to identify similar drugs, which can be potential candidates for drug repurposing application. The main contributions of this paper include:

(1) We use a generative model for synthesis of drug-based high-content imaging data.
(2) We use a clustering method to evaluate the quality of synthetic images by obtaining similarity scores between the cluster-based drug signatures computed for real and synthetic images.
(3) We use the synthetic images to augment the real imaging data. The use of augmented dataset is shown to improve the quality of clusters.
(4) Finally, we extract drug similarity scores from the computed signatures, which are used to identify phenotypically similar drugs. We also show that the confidence in these drug similarity scores improves with the use of augmented dataset for clustering.

2 Methods

2.1 Overview of the Framework

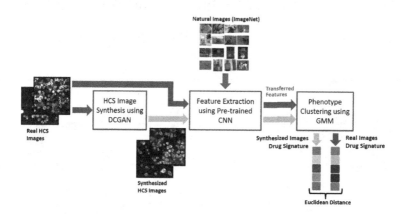

Fig. 1. Overview of the framework for generation of high-content image data

The proposed framework for the generation of high-content cellular image data which can be used for phenotypic analysis of drug-induced effects on cells is shown in Fig. 1. This framework comprises of three main modules, namely HCS image generation module, feature extraction module and phenotypic clustering module. The images collected by conducting the HCS experiments form the input to the image generation module. Next, a pre-trained CNN model is applied to extract features for real as well as synthesized images. Finally, an unsupervised clustering module is used to generate drug signatures and similarities between these signatures. This similarity score is used to evaluate the quality of generated images by comparing signatures of a particular drug based on real and synthetic images. These modules are presented in detail in the next section.

2.2 HCS Imaging Data

Human monocyte THP-1 cells were treated with different FDA approved drugs for 72 hrs in a 384-well format HCS assay. These cells were then harvested and stained with Phalloidin (which stains the actin cytoskeleton present in cell membrane). The assay also contained control wells with cells treated with vehicle control (DMSO) which is not expected to have any effect on the cells. Images of the stained cells were acquired at 40X magnification from the cell membrane using the high-content imaging system IN Cell Analyzer 2000. Each image is an RGB image with the resolution of 2048 × 2048 pixels, where the pixel size was $0.185 \times 0.185\,\mu m^2$. The developed framework for HCS image data synthesis was applied on a drug assay plate containing multiple FDA approved drugs. The assay consisted of 5 well replicates for each drug with 36 fields × 4 images per field = 144 images per well, resulting in 720 images acquired from cell membrane channel for each drug treatment.

2.3 Generative Modeling: HCS Image Synthesis

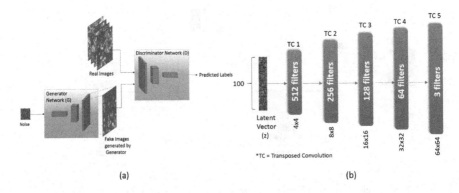

Fig. 2. (a) Architecture of DCGAN. (b) Generator Network.

In this work, we use one of the most popular variants of GAN - DCGAN, for HCS imaging data generation [13]. The architecture of the DCGAN employed in this work for HCS image generation is showed in Fig. 2(a). As can be seen from the figure, it consist of two networks, the discriminator network D and the generator network G. Here, both generator and the discriminator are CNN architectures. The generator network takes input as noise and attempts to generate fake samples such that these samples will be considered as original samples with high probability. The generator generates samples $s_1, s_2, ..., s_m$ from a uniform distribution u. The final goal of the generator is to generate samples such that distribution of both the generated as well as original samples are same. The discriminator on the other hand takes the original samples $x_1, x_2, ..., x_m$ as well as the samples generated by the generator $s_1, s_2, ..., s_m$ and tries to differentiate among them. The generator and discriminator network are trained simultaneously. These networks are trained using the following loss function as two player game-theoretic optimization problem:

$$\min_G \max_D \mathbb{E}_{x \sim l_{data}} \, log D(x) + \mathbb{E}_{s \sim l_s}[log(1 - D(G(s)))] \tag{1}$$

As a part of the training routine the generator improves its ability to generate samples which are similar to the real ones while the discriminator evolves such that it has the ability to successfully discriminate the real and the fake samples.

Generator Architecture: The generator architecture is shown in Fig. 2(b) (adapted from [13]). It is a deep CNN architecture. The generator takes as input a vector of uniform distribution and size 100 and finally outputs an 3 channel image of cell of 64 height and width. First, the input vector is reshaped and a transposed convolution operation is performed on it. This convolution layer has a kernel size of 4×4 and 512 filters. Further, different filters are applied subsequently as shown in Fig. 2(b). These filters have a kernel size of 4×4 in every layer. The generator architecture consist of five transposed convolution layers. All the layer employ batch normalization except the last layer. Normalizing the data to zero mean and unit variances over all the batches of data stabilizes the training and helps in convergence. In the final layer a *Tanh* activation function is employed. The rest all the other layers use the popular *ReLU* as the activation function.

Discriminator Architecture: Similar to the generator architecture, the discriminator also employs a deep convolutional neural network. In the Fig. 2(b), we only show the generator network for the sake of simplicity. The discriminator network takes an image of size $64 \times 64 \times 3$ as input and predicts whether the sample is *fake* or *real*. Similar to the generator network, the discriminator network has five convolution layers, each with a kernel size of 4×4. The convolutions in each steps help reduce the size of images. The network employs leaky ReLU in all the layers and *Sigmoid* activation function in the last layer for binary classification. Similar to generator, batch normalization is also employed in each of the layers expect the last layer.

DCGAN Hyperparameters and Training Procedure: We follow an iterative procedure for training the generator and discriminator. The data was divided into mini batches of size 128. For preprocessing, we scale the images in the range [0, 1]. The slope of the leaky ReLU was set to 0.2 and the weights of the networks are initialized using a normal distribution of zero mean and standard deviation with 0.02. We employed the mini-batch stochastic gradient descent with Adam optimizer [7] for learning the weights of the networks. The learning rate employed for the training is 0.0002. The training procedure is carried out for 500 epochs.

We have also employed other generative modeling approaches to synthesize HCS images and compared the quality of images as well as training time of all these methods with that of DCGAN. First, we used a basic variant of GAN (Vanilla GAN) [4] which uses fully connected layers followed by Leaky ReLU nonlinearity in both generator and discriminator, with Tanh and sigmoid activation functions used in the last layer of generator and discriminator respectively. Next, we studied one of the state-of-the-art variants of GAN, Progressive GAN (ProGAN), which was adapted from [6]. For ProGAN training, the generator (G) and discriminator (D) grow incrementally, i.e. initially G and D learn from a low resolution of 4 × 4, and layers are added into the G and D progressively as the training advances. This is done until we attain the desired resolution of the HCS image of size 64 × 64. We also implemented a Convolutional Variational Autoencoder (CVAE) model [8] for comparison with the proposed DCGAN architecture. The model consists of an encoder and a decoder in which the decoder is the generator network and the encoder is formed by four successive convolution layers, which are symmetric to the four deconvolution layers in the decoder. After the model is trained, the encoder returns the means and variances of 100 learned Gaussian distributions. To generate a new image, we randomly draw a 100-component vector from the learned distributions and pass it to the decoder. We trained our CVAE model for 500 epochs using Adam optimizer, learning rate of 0.001 and batch size of 32. The loss function used is the sum of reconstruction loss (binary cross entropy) with KL divergence.

2.4   Pre-trained CNN Based Feature Extraction

In this work, the CNN model has been applied to extract features from both real as well synthetic HCS images. CNN's are a class of deep learning which employ operations such as filtering and pooling at various levels to learn feature representation of the images. In recent years, CNN has achieved excellent results in different fields such as computer vision and image recognition [5]. The generated features by various CNN architectures are able to learn the various levels of representations for the data sets. Many works in the literature rely on pre-trained models to transfer features, because of lesser number of samples [15]. In this work, we employed the well known CNN architecture VGG-16 for feature extraction [14]. The VGG network was trained on the popular ImageNet dataset consisting of 13 million images. The neural network was used for feature extraction by first resizing the images from 2048 × 2048 pixels to 64 × 64 pixels and then passing the resized images through its pre-trained version without fine tuning, such that the final classification layer of the network was cut off and the penultimate layer represented feature embedding. In particular, the VGG architecture uses convolutional layers with very small receptive fields (3×3), stacked on top of each other in increasing depth, and max pooling layers of size 2×2, followed by three fully-connected layers. Rectification nonlinearity (ReLU) activation is applied to all hidden layers. Here, "16" stands for the number of hidden layers in the network.

2.5 GMM Based Phenotype Clustering

The phenotypic similarities between different drug effects were determined using unsupervised clustering method based on fitting GMM [1] to the cellular image features extracted using pre-trained CNN for both real and synthetic images. The parameters of GMM were estimated using the Expectation-Maximization (EM) algorithm [3]. The number of clusters (K) in the data was evaluated by computing the Bayesian Information Criterion (BIC), which attempts to maximize the log likelihood of the data samples for the given model parameters while minimizing the model complexity to avoid overfitting. Hence, the number of clusters (K) for the GMM was chosen as the one with the smallest BIC value over different values of K ranging from 2 to 20. In order to identify groups of drugs with similar cellular effects, we first defined drug signatures based on the clustering results. These cluster signatures were generated by computing the fractions of drug-treated images falling into different clusters. Finally, the similarity measure between two drugs was derived by computing the Euclidean distance between their cluster signatures.

3   Results

In this section, we present the results obtained when our proposed HCS image generation pipeline was applied to the image data collected from a drug screen. The image data from the drug screen comprised HCS images corresponding to 1) cell-DMSO: control cells treated with DMSO, and 2) cell-Dx: cells treated with a drug Dx. Our analysis will focus on six drugs screened in this assay, which are Moxalactam, Ganciclovir, Rabeprazole, Isoniazid, Metformin and Streptomycin. The image dataset consisting of 720 images per drug treatment was divided into training and test sets by fixing a test dataset containing 1 replicate of each drug, which corresponds to 144 images per drug. The training dataset was then used for generating images using DCGAN, which were combined with the real training images to yield an augmented HCS dataset.

Firstly, we present the results pertaining to synthesis of cellular images which are used for data augmentation and evaluating the quality of drug phenotype clusters. Secondly, we analyze the similarity between drug signatures of real and synthetic images, and study the drug-specific signatures. Finally, we analyze the distance between drug signatures to identify similar drugs and use the augmented data to improve the similarity scores.

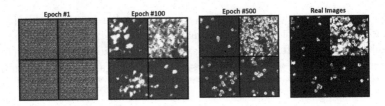

Fig. 3. Progressive learning of cell images over different epochs

3.1   Generation of HCS Images: Comparison of Generative Models

Figure 3 shows the progression of 2×2 cell-Metformin images synthesized after different number of epochs, i.e. after 1, 100 and 500 epochs. It can be seen that the synthesized cell images progress from images of random noise at the beginning of DCGAN training (epoch #1) to images of small bright blobs (epoch #100) and finally start looking very similar to real cell images (right panel) with appearance of cell membranes at epoch #500. Next, we compared the performance of DCGAN with other generative models (Vanilla GAN, ProGAN and CVAE) by studying the quality of synthetic images and the training time required in generating the HCS images. Figure 4(a) shows the generated synthetic images and the real images (green border). It can be seen that the images generated by DCGAN and ProGAN are close to the real images, however, the images generated using CVAE are not completely realistic with cell membranes not clearly visible, and finally Vanilla GAN generated images looking very different. The image feature distributions (Gaussian density estimate of the first principal component (PC) of features) of the generated images in Fig. 4(b) also show how the synthesized image distributions of DCGAN and ProGAN generated images are close to that of real images (black), with Vanilla GAN image distribution farthest apart. Next, we looked at the training times taken by these methods for HCS image generation. Figure 4(c) shows the distance between real and synthetic image feature distributions, computed as the Kolmogorov-Smirnov statistic between the real and synthetic image features transformed into PCs, as a function of training time in minutes. It can be seen that even though HCS images generated by DCGAN and ProGAN are similar and close to real images, ProGAN takes much longer time (≈ 1000 min) compared to DCGAN (≈ 30 min), while CVAE has still not converged even after ≈ 2000 min. Hence, both DCGAN and ProGAN can be employed for HCS image generation, however, DCGAN generates realistic images in much lesser time making it the candidate of choice in our proposed framework.

3.2   Clustering Performance: Evaluation of Synthetic Images

Next, we studied the effect of combining these synthesized images with the real images on the quality of clusters generated using augmented dataset. GMM based clustering was done on training dataset comprising only real images, which

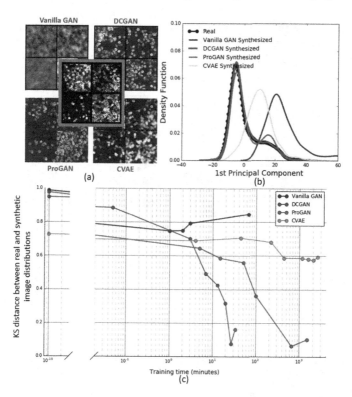

Fig. 4. Comparison of different generative models for HCS image generation. (a) Synthesized images vs real images (green border). (b) Synthesized image feature distributions compared with real image feature distribution (black). (c) Distance of synthesized image distributions from real image distribution as a function of training time (minutes) for different methods. (Color figure online)

resulted in 7 clusters. Figure 5(a) shows PCA scatter plot of the image samples of Streptomycin and Metformin drugs falling in these 7 clusters, and the drug signatures computed from the number of drug samples grouped in different clusters (heatmaps at the right). The drug signatures were also derived for the test images based on the clusters learned on training images. Euclidean distance between the training and test drug signatures was used as a measure to evaluate the quality of clusters as well as the quality of synthesized images, with smaller distances indicating similar training and test signatures for a particular drug, and hence clusters generalizing well to test dataset.

Figure 5(b) shows the mean of distances between cluster signatures computed for training and test images for all six drugs as a function of number of real and synthesized images used for training. As increasing number of synthetic images were added to real images to generate the drug clusters, the distance between training data and test data signatures became smaller. This is most evident for the case when different numbers of synthesized images were combined with 100 real images per drug (Fig. 5(b), blue). Hence, we can conclude that the synthetic images are realistic, which can contribute to improved clustering performance, when combined with real images.

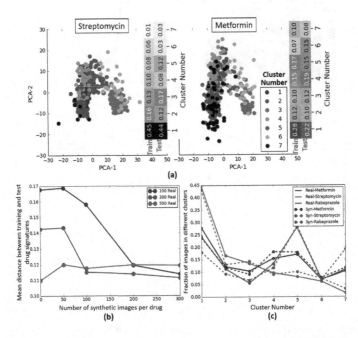

Fig. 5. (a) Real image samples of two drugs, as grouped into 7 GMM clusters. Signatures computed for the training and test samples. (b) Distance between the drug signatures for training and test images decreases with increasing number of synthetic images added to real images (100, 200, 300 per drug) for training. (c) Signatures of real (Real-Dx, solid) and synthetic (Syn-Dx, dashed) images of 3 drugs. (Color figure online)

We have also evaluated the quality of synthesized images by studying how well the clusters learned on real images transfer to synthesized images of different drugs. Figure 5(c) shows signatures of three drugs based on the 7 clusters obtained from training on only real images. The signatures for these three drugs look different (Metformin and Rabeprazole more similar to each other than to Streptomycin) and moreover, for a particular drug, the real images (Real-Dx) and synthesized images (Syn-Dx) have very similar signatures. This suggests that the HCS cellular images generated by DCGAN are realistic and also capture the drug-specific cellular effects. Hence, this kind of analysis can help to discover drugs having similar phenotypic effects, potentially for drug repurposing applications, wherein the HCS image generation/augmentation can alleviate the requirement of large amount of image data. This is discussed further in the next section.

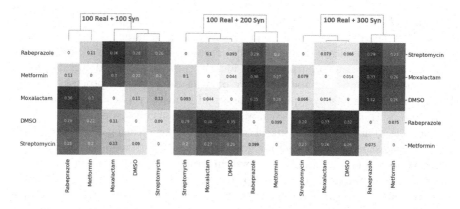

Fig. 6. Distance maps of 5 drugs showing 2 groups of similar drugs. The similarity scores within these 2 groups improve as more number of synthetic images (left-100, middle-200, right-300 per drug) are added to 100 real images per drug.

3.3 Grouping of Similar Drugs: Improved Confidence

Finally, we studied how creating an augmented HCS image dataset can help to improve the confidence in drug similarity scores (Euclidean distance between their signatures) derived from the GMM-based phenotypic clusters of these drugs. Figure 6 shows the distance maps for four different drugs and one control (DMSO), as more number of synthetic images (100, 200 and 300) are combined with a limited number (100 per drug) of real images to perform phenotypic clustering. As shown in the distance map on the left (100 Real + 100 Syn), two groups of drugs are suggested, with Rabeprazole and Metformin in one group and Moxalactam and Streptomycin in the second control-like group. As more number of synthetic images (200 per drug, middle) were added, the drug similarity scores within these two groups improved. With further augmentation of the real data (300 Syn images, right), the similarity scores improved further i.e. distance between Rabeprazole and Metformin signatures reduced from 0.11 to 0.075 and the distances between control-like drugs also reduced with addition of more synthetic images to limited number of real images. Hence, the confidence in our drug similarity results improved with the use of augmented HCS dataset created from synthetic images generated by DCGAN.

4 Discussion

In this paper, we proposed a DCGAN-based framework for generating cellular images from high-content images collected in a drug screening assay. We demonstrated the application of this framework on synthesizing realistic drug-treated images of cell populations to improve the confidence in predicting phenotypically similar drugs. This involved training the GMM model on augmented dataset consisting of real and synthesized images. This kind of approach of synthesizing realistic cellular images can enable generation of large volumes of image data, which can facilitate the use of deep learning algorithms for drug discovery applications, like drug repurposing to identify new uses for existing approved drugs, thereby improving the efficiency of drug development process. A high-throughput imaging assay typically consists of multiple replicates for each drug compound, with one assay screened at a time to focus on a specific disease or a condition. A generative modeling based approach for creating augmented HCS datasets consisting of fewer real images and higher number of synthetic images can allow to reduce the number of experimental replicates for each compound screened. This approach can also help to improve the efficiency of HCS experiments by screening more compounds as well as screening compounds for multiple diseases/conditions in the same assay, leading to reduced experimental costs.

References

1. Bishop, C.M.: Pattern Recognition and Machine Learning. Springer, New York (2006)
2. Breinig, M., Klein, F.A., Huber, W., Boutros, M.: A chemical-genetic interaction map of small molecules using high-throughput imaging in cancer cells. Mol. Syst. Biol. **11**(12), 846 (2015)
3. Dempster, A.P., Laird, N.M., Rubin, D.B.: Maximum likelihood from incomplete data via the EM algorithm. J. Royal Stat. Soc. Ser. B (Methodol.) **39**(1), 1–22 (1977)
4. Goodfellow, I., et al.: Generative adversarial nets. In: Advances in Neural Information Processing Systems, pp. 2672–2680 (2014)
5. He, K., Zhang, X., Ren, S., Sun, J.: Deep residual learning for image recognition. In: Proceedings of the IEEE Conference on Computer Vision and Pattern Recognition, pp. 770–778 (2016)
6. Karras, T., Aila, T., Laine, S., Lehtinen, J.: Progressive growing of GANs for improved quality, stability, and variation. arXiv preprint arXiv:1710.10196 (2017)
7. Kingma, D.P., Ba, J.: Adam: a method for stochastic optimization. arXiv preprint arXiv:1412.6980 (2014)
8. Kingma, D.P., Welling, M.: Auto-encoding variational Bayes. arXiv preprint arXiv:1312.6114 (2013)
9. Kraus, O.Z., Ba, J., Frey, B.J.: Classifying and segmenting microscopy images with deep multiple instance learning. In: Bioinformatics (2016)
10. Kraus, O.Z., et al.: Automated analysis of high-content microscopy data with deep learning. Mol. Syst. Biol. **13**(4), 924 (2017)
11. LeCun, Y., Bengio, Y., Hinton, G.: Deep learning. Nature **521**(7553), 436–444 (2015). https://doi.org/10.1038/nature14539
12. Putin, E., et al.: Reinforced adversarial neural computer for de novo molecular design. J. Chem. Inf. Model. **58**(6), 1194–1204 (2018)
13. Radford, A., Metz, L., Chintala, S.: Unsupervised representation learning with deep convolutional generative adversarial networks. arXiv preprint arXiv:1511.06434 (2015)

14. Simonyan, K., Zisserman, A.: Very deep convolutional networks for large-scale image recognition (2014)
15. Yosinski, J., Clune, J., Bengio, Y., Lipson, H.: How transferable are features in deep neural networks? In: Advances in Neural Information Processing Systems, pp. 3320–3328 (2014)
16. Young, D.W., et al.: Integrating high-content screening and ligand-target prediction to identify mechanism of action. Nat. Chem. Biol. 4(1), 59 (2008)

First Pacific Asia Workshop on Game Intelligence and Informatics (GII 2020)

A Study of Game Payment Data Mining: Predicting High-Value Users for MMORPGs

Junxiang Jiang[✉]

FinVolution Group, Shanghai, China
jx.jiang@outlook.com

Abstract. A user is high-value when his payment amount is more than a predetermined amount (such as 100 dollars). We first analyze payment data under the character level (CLV) dimension of six real mobile MMORPGs and verify the importance of high-value users for achieving game profit and the necessity of HU prediction. Afterward, we propose a CLV-based high-value user prediction (CLVHUP) model that solves limitations of the existing method (delayed identification, the uncertainty of the predetermined amount, and the inflexible time window). This model not only can predict high-value users with sparse features and a small volume of users but is also sensitive to the information underlying CLVs. To the best of our knowledge, this is the first work to analyze payment behavior under the CLV dimension and to predict high-value users by utilizing machine learning techniques and exploring the information underlying CLV. Comprehensive experimental results demonstrate that the performance of our solution is better than that of several baselines.

1 Introduction

The Ministry of Industry and Information Technology of China reported that the profit of Chinese online games reached 129.4 billion yuan between January and August 2018, which increased by 24.8% year-to-year and made up 21.73% of the profit of chinese Internet companies[1]. Massive multiplayer online role-playing games (MMORPGs) are an important part of online games, and the revenue comes primarily from user payments. With the development of the online game market, people pay increasing attention to MMORPGs.

In MMORPGs, a user is high-value when his payment amount is more than a predetermined amount (such as 100 dollars). We first analyze payment data under the character level (CLV) dimension of six real mobile MMORPGs and verify the importance of high-value users for achieving game profit and the necessity of HU prediction. The existing method identifies a user as a high-value user when the user's payment amount is greater than a predetermined amount during a fixed time window. However, this method has limitations: (1) Delayed identification. Only if users spent enough money during a period do operators identify

---

[1] http://www.miit.gov.cn/n1146312/n1146904/n1648355/c6407186/content.html.

W. Lu and K. Q. Zhu (Eds.): PAKDD 2020 Workshops, LNAI 12237, pp. 181–192, 2020.
https://doi.org/10.1007/978-3-030-60470-7_17

them as high-value users. (2) The uncertainty of the predetermined amount. A higher predetermined amount loses many high-value users, while a lower predetermined amount cannot identify high-value users effectively. (3) Inflexible time windows. Payment behavior is strongly associated with payment points that are set at each character level (CLV) and not a specific time window.

Incorporating models (machine learning methods) with payment behavior features underlying CLVs (payment behaviors are generally associated with the CLVs and are the most related to our target) can solve the aforementioned methods' limitations. For delayed identification, operators can identify high-value users when their payment amounts are less than the predetermined amount. For the uncertainty of the predetermined amount, instead of selecting users according to the predetermined amount, operators can select any number of users after ranking users in descending order of model outputs. For the inflexible time window, models use CLVs to replace the fixed time window for measuring payment behavior. However, models also have three challenges: (1) sparse features (users usually pay for games in several CLVs among dozens of hundreds of CLVs); (2) a small volume of users (only tens of thousands of paid users); (3) insensitivity to the information underlying CLVs. Therefore, we propose a new CLV-based high-value user prediction (CLVHUP) model.

In summary, our contributions are as follows.

- We quantitatively analyze the payment behavior under the CLV dimension and verify the importance of high-value users for achieving game profit and the necessity of HU prediction.
- We first utilize machine learning methods to predict HUs in MMORPGs. To capture the information underlying CLVs, we propose the CLVHUP model.
- In a series of experiments, we verified the effectiveness of CLVHUP by comparing it with LR, SVM, FM and MLP.

The rest of this paper is organized as follows. Section 2 introduces the datasets used in the data analysis and experiments. Section 3 presents our data analysis. Section 4 introduces CLVHUP in detail. In Sect. 5, we perform a set of experiments and comparisons by using the payment data of six MMORPGs. Section 6 introduces related works. Finally, Sect. 7 concludes the paper with future work.

2   Datasets

Our datasets are composed of the payment data of six real mobile MMORPGs and provided by a chinese mobile game development and publication start-up. Since the lifecycle of mobile games is very short (approximately half or one year), identifying high-value users earlier is much more important for game companies. We select more than three months of payment data for analysis and experiments. We do not disclose the payment data detail and game names because of business confidentially. Instead, we use numbers 15, 59, 98, 108, 111 and 163 to represent these six MMORPGs. The maximum character levels of these six MMORPGs are 91, 133, 97, 112, 100 and 143, respectively. The number of payment users in each game is more than 40 thousand. In the data analysis section, we use the real predetermined number. In the experiments and results section, we decrease the predetermined number to increase the proportion of high-value users. The proportions of high-value users in each game are 30.75%, 39.75%, 38.37%, 24.83%, 25.57% and 39.23%.

3   Data Analysis

3.1   Preliminary

Operators identify high-value users according to the user payment amount during a fixed time window. In this section, we not only reveal the weaknesses of the operator used method but also support our work from a data perspective. For clarity, our data analysis answers (A) the following questions (Q).

- Q 1: *How important are high-value users in the perspective of game income?*
- Q 2: *How important are high-value users in the perspective of improving the HU number?*
- Q 3: *How important is HU prediction in the perspective of the payment environment?*
- Q 4: *How important is HU prediction in the perspective of payment amount?*
- Q 5: *How many potential high-value users will convert to high-value users?*

To facilitate expression, we use **high-CLV users** to represent the top 30% CLV users, **mid-CLV users** to represent the top 60% to 30% CLV users, and **low-CLV users** to represent the rest of the users.

3.2   Analysis Payment Behavior Underlying CLVs

In this section, we analyze payment behavior with the change in CLVs. For easy analysis, we divide CLVs into several intervals using the calculation below.

$$interval = \left(\left\lfloor \frac{CLV}{10} \right\rfloor + 1 \right) \tag{1}$$

where *interval* refers to the character level interval and CLV refers to the CLV.

Figure 1 analyzes the proportion of high-value users to paid users. Figure 2 analyzes paid user behavior at different CLV intervals. We exclude the data of the first 10 CLVs in Fig. 2 because users receive one-time welfare (such as spending 1 dollar to buy higher-value goods) when they begin playing the game.

A 1: By analyzing Fig. 1, we find that the proportion of high-value users to paid users is less than 20% when CLV is no more than 30. In general, the proportion of high-value users in the low-CLVs is less than 20%. With the increase in the CLV, the proportion of high-value users (except for the No. 111 game) continuously increases, and finally more than 70%. These analyses reveal that high-value users play an important role in profit pursuit; not only do high-value users contribute a great deal of profit at mid-CLVs, but also the majority of paid users at high-CLVs are high-value users. To achieve an increase in profit, to identify potential high-value users and further converting them to high-value users is a good method.

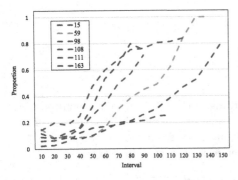

Fig. 1. Proportions of high-value users at different character level intervals

Table 1. Proportion (%) of potential high-value users at different CLV intervals.

Interval	Games					
	15	59	98	108	111	163
1–40	16.94	12.00	16.40	14.39	22.73	59.16
41–80	00.12	02.03	00.19	00.52	03.54	21.32
81–120	00.00	00.00	00.00	00.00	00.00	02.11
121–160	/	00.00	/	/	00.00	00.00

***A* 2:** We analyze the number of paid users at different CLV intervals, as shown in Fig. 2*. The number of paid users continually declines with increasing CLVs. There are three reasons for this phenomenon. First, higher-CLV users are less common since CLV upgrading becomes increasingly difficult. Second, users become skillful after spending more time playing the game, so they do not need to pay to enjoy the game. Third, many paid users are lost. To improve the number of paid users, it is a good choice for operators to convert low-CLV paid users to high-value users because the population of low-CLV paid users is larger in contrast to high-CLV paid users, and they will encounter more payment points in the future. In practice, the proportion of paid users is much larger than the operators, therefore, accurately identifying paid users who will be high-value becomes more important.

***A* 3:** We analyze the payment frequency at different CLV intervals, as illustrated in Fig. 2†. First, on average, the number of payment behaviors continually declines when the CLV increases. The payment behavior of high-CLV paid users is rarer compared with low-CLV paid users, which reveals that high-CLV paid users cannot fully activate the payment environment. Second, we find that the trends of the curves in Fig. 2* and Fig. 2† are approximate, which means that the higher the popularity is, the larger the payment number is. This phenomenon further demonstrates that low-CLV and mid-CLV paid users are the basis for activating the payment environment. Third, the number of dominant low-CLV paid users reflects that low-CLV paid users play a key role in profit increase.

***A* 4:** We analyze Fig. 2‡, which shows the payment amount at different CLV intervals. First, the payment amount continually declines with an increasing CLV, in general. Paid users at low- and mid-CLV provide more profit from

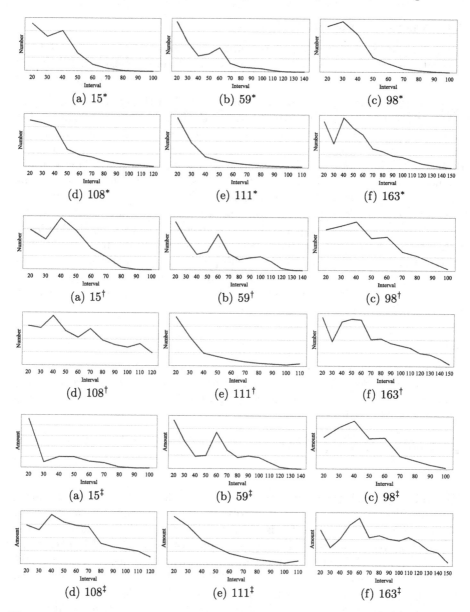

Fig. 2. Analysis of paid users at different CLV intervals (∗ refers to the number of paid users at different CLV intervals, † refers to the number of payment behaviors at different CLV intervals, ‡ refers to the payment amount in different CLV intervals.)

games. Second, compared to Fig. 1, we find that half of the paid users at low- and mid-levels are not high-value users. It is helpful to maintain the high desire payment of these users to increase game profit, considering that these users will encounter many payment points in the future.

3.3   Analysis of Potential High-Value Users

Table 1 shows the proportion of potential high-value users at different CLV intervals. We divide every 40 CLVs into a CLV interval considering that the data scale within 40 CLVs is suitable for training a model in the experiments.

A 5: In the interval of 1 to 40, more than 12.00% of the high-value users are transformed from normal paid users. In particular, 59.16% of the high-value users are transformed from normal paid users in the No 163 game. Identifying high-value users is useless when their CLVs are larger than 80, which reflects the importance of setting a suitable time window in the existing method and the necessity of predicting HU when their CLVs are lower. The potential high-value users in the interval of 1 to 40 have an absolute quantitative advantage compared with other intervals. Identifying potential high-value users when the CLVs of users are between 1 and 40 and converting them to high-value users is helpful for improving the user experience and the profit from the games. Furthermore, Table 1 only illustrates the proportion of high-value users that are naturally converted from potential high-value users. In consideration of user churn and the help of operator services, there exist more potential high-value users in reality.

3.4   Summary

In conclusion, high-value users are important for achieving game profit. In addition, converting low- and mid-CLV users to high-value users is necessary for improving the user experience and gaming profit, and this finding verifies the necessity of HU prediction.

4   Proposed Model

4.1   Notations

We utilize the users payment behaviors $\mathcal{X}$ to predict whether their are high-value $\mathcal{Y}$. We use $x = \{x_i\}_{i=1:n} \in \mathcal{X}$ to represent the feature vector of a user. The length of the feature vector is equal to n, and x_i is the i-th feature of the user. We divide the feature vector by CLV and obtain a set of CLV features $\mathcal{D}$. $d = \{d_i\}_{i=1:l} \in \mathcal{D}$ is used to represent the CLV features of a user, l is the number of CLVs. d_i represents the feature vector of the i-th CLV for the user, the length is equal to k ($k = \frac{n}{l}$). We treat this problem as a binary classification problem, and thus, we use $y \in \{0,1\} \in \mathcal{Y}$ to denote whether the user is high-value. If y is equal to 0, the user is not high-value; otherwise, the user is high-value. Our goal is to learn a prediction function f on a training set $\mathcal{T} = \{y \mid x\}$ where y is known, and to use the function f to predict the most likely label for each user on the test set $\mathcal{T}' = \{y' \mid x'\}$ where y' is unknown. We formulate the HU prediction task as follows:

$$f(x) \rightarrow y^* \tag{2}$$

where y^* is the predicted value. We optimize the parameters of f by minimizing the differences $|y - y^*|$ between the predicted value and the real value over $\mathcal{T}$.

4.2   Models

We use the following equation to predict the probability that a user is high-value.

$$y(x) = \frac{1}{1 + exp\left(-\sum_{i=1}^{l} level_i\right)} \tag{3}$$

where $level_i$ refers to the impact of the i-th CLV on the result. To better capture the characteristic of payment behavior, we introduce a group of vectors $\mathcal{V} = \{v_i\}_{i=1:l}$. v_i (the length is equal to k) represents the characteristic of the i-th CLV. Therefore, the impact of each CLV can be divided into the low level part and the high level part.

$$level_i(x) = \underbrace{w_0 + W_i^\top d_i}_{low} + \underbrace{d_i^\top M_i v_i}_{high} \tag{4}$$

In the low level part, $\mathcal{W} = \{W_i\}_{i=1:l} = \{w_i\}_{i=0:n}$ captures the impact of behavior at each CLV, and the length of W_i is equal to k. In the high level part, $\mathcal{M} = \{M_i\}_{i=1:l}$ captures the relation between behavior and the characteristic of the CLV, M_i is a $k \times k$ matrix. CLV is a special time series; payment behaviors in CLVs occur in turn, and behaviors in prior CLVs are not affected by behaviors in posterior CLVs. Payment behaviors are related to each other: first, prior behaviors influence posterior behaviors; second, posterior behaviors can be used to infer prior behaviors. Similarly, we can use this pattern to deeply mine the information underlying CLVs since payment behaviors are associated with CLVs. Therefore, the CLVHUP formula is as follows:

$$p(x) = \text{sigmoid}\left(\sum_{i=1}^{l} \left(w_0 + W_i^\top d_i + d_i^\top M_i v_i\right)\right)$$

$$= \text{sigmoid}\left(w_0 + \sum_{i=1}^{n} w_i^\top x_i + \sum_{i=1}^{l} \left(d_i^\top M_i v_i\right)\right) \tag{5}$$

$$\text{s.t.} \quad v_i = \sum_{j=1,j\neq i} d_i^\top H_j v_j$$

where $H_j \in \mathcal{H} = \{H_i\}_{i=1:l}$ represents relation between current CLV and j-th CLV. $H_i = \{h_{ij}\}_{j=1:k}$ is a vector, and the length is equal to k.

4.3   Model Inference

Because HU prediction is treated as a binary classification problem, we use maximum likelihood estimation to construct the objective function. We use stochastic gradient descent (SGD) to optimize the parameters. The likelihood is as follows:

$$\mathcal{L}(x, y) = p(x)^y (1 - p(x))^{1-y} \tag{6}$$

We add regularization to the log-likelihood to obtain the following optimization problem:

$$
\begin{aligned}
\mathcal{J}(x,y) &= \log \mathcal{L}(x,y) + \Omega(\mathcal{V}, \mathcal{M}, \mathcal{W}) \\
&= y \log p(x) + (1-y) \log(1 - p(x)) + \Omega(w_0, \mathcal{W}, \mathcal{M}, \mathcal{V}, \mathcal{H})
\end{aligned}
\tag{7}
$$

Here, Ω is the regularization term, and we adopt the Frobenius norm to avoid overfitting.

$$
\Omega(w_0, \mathcal{W}, \mathcal{M}, \mathcal{V}, \mathcal{H}) = \frac{\lambda}{2}(\|w_0\|_F^2 + \|\mathcal{W}\|_F^2 + \|\mathcal{M}\|_F^2 + \|\mathcal{V}\|_F^2 + \|\mathcal{H}\|_F^2)
\tag{8}
$$

Our target is to maximize the objective function $\mathcal{J}(x,y)$ by finding suitable parameters.

$$
\arg\max_{\theta} \mathcal{J}(x,y)
\tag{9}
$$

where θ represents parameters that need to learn.

We provide the generalized gradients of parameters which are illustrated as follows:

$$
\frac{\partial \mathcal{J}}{\partial \theta} = (p(x) - y)\frac{\partial p(x)}{\partial \theta} + \frac{\partial \Omega(\theta)}{\partial \theta}
\tag{10}
$$

After we add the constraint of v_i, the likelihood function is transformed and shown as below:

$$
p(x) = \text{sigmoid}\,(p'(x))
\tag{11}
$$

$$
\begin{aligned}
p'(x) &= w_0 + \sum_{i=1}^{n} w_i^\top x_i + \sum_{j=1, j\neq i}^{l} (d_j^\top M_j v_j) + d_i^\top M_i v_i \\
&= w_0 + \sum_{i=1}^{n} w_i^\top x_i + \sum_{j=1, j\neq i}^{l} \left(d_j^\top M_j \sum_{g=1, g\neq j}^{l} d_g^\top H_g v_g \right) + d_i^\top M_i v_i
\end{aligned}
\tag{12}
$$

The $\frac{\partial p'(x)}{\partial \theta}$ is shown as below, and the gradient of $\frac{\partial \Omega(\theta)}{\partial \theta}$ is equal to $\lambda\theta$.

$$
\frac{\partial p'(x)}{\partial \theta} =
\begin{cases}
1, & \text{if } \theta \text{ is } w_0 \\
x_i & \text{if } \theta \text{ is } w_i \\
d_i v_i & \text{if } \theta \text{ is } M_i \\
M_i^\top x_i + H_i^\top d_i \sum_{j=1, j\neq i}^{l} M_j^\top d_j & \text{if } \theta \text{ is } v_i \\
d_i v_i \sum_{j=1, j\neq i}^{l} d_j M_j & \text{if } \theta \text{ is } H_i
\end{cases}
\tag{13}
$$

5   Experiments and Results

5.1   Experimental Settings

Data Processing. We select paid users who do not reach the high-value criterion before the 40-th CLV as the dataset. A paid user who finally becomes a high-value user is a positive sample; otherwise, it is a negative sample. We randomly select 80% of the dataset as the training set and the rest as the test set. We randomly select 10% of the training set as the violation set and use it to find suitable hyperparameters (e.g., learning rate). We observe that the proportions of positive and negative samples in the training set are unbalanced (roughly, positive samples: negative sample = 1:20), so we copy the positive samples until the positive and negative samples are balanced. To reduce experimental bias, we calculated the average value of 10 experimental results.

Baselines. To verify the effectiveness of the proposed model, we compare it with four baselines: (1) logistic regression (LR) with L2 regularization, and we use SGD to optimize the parameters; (2) support vector machine (SVM) with Gaussian kernel; (3) we compare with a factorization machine (FM) [3], and we use the alternating least squares (ALS) [4] method to optimize the parameters; (4) multiple layer perceptron (MLP) is a 3-layer fully connected network.

Metrics. We use four metrics to measure the performance of the models: area under the curve (AUC), Recall, F1, F3 and F5. AUC measures the performance of the classification. Operators use models to identify high-value users in practice after seting a fixed classification threshold for all games. And they are more concerned about the number of identified high-value users, so we adopt metrics of recall, F1, F3 and F5. Please note that the higher the value of the metrics, the better the performance of the models.

Table 2. Experimental results in AUC.

Games	Models				
	LR	SVM	MLP	FM	CLVHUP
15	0.6947 ± 0.0338	0.664 ± 0.0294	0.754 ± 0.0392	0.7878 ± 0.0211	**0.8169 ± 0.0186**
59	0.5931 ± 0.0159	0.5359 ± 0.0084	0.6211 ± 0.0151	0.6231 ± 0.0186	**0.6361 ± 0.0209**
98	0.6899 ± 0.0223	0.6424 ± 0.0199	0.6763 ± 0.0255	0.7445 ± 0.0203	**0.7636 ± 0.0175**
108	0.6968 ± 0.0509	0.5709 ± 0.0426	0.6878 ± 0.0589	0.8276 ± 0.0271	**0.8325 ± 0.0263**
111	0.6706 ± 0.0283	0.6236 ± 0.0316	0.6911 ± 0.0302	0.7653 ± 0.0123	**0.7876 ± 0.0128**
163	0.6189 ± 0.0126	0.6076 ± 0.0101	0.6849 ± 0.0092	0.7114 ± 0.0063	**0.7226 ± 0.0063**
Avg	0.6607	0.6074	0.6858	0.7433	**0.7599**

5.2   Performance

In this section, we verify the effectiveness of the proposed model by analyzing the experimental results. Table 2 illustrates the experimental results of the models in AUC. Table 3 demonstrates the performance of the models in other metrics (recall, F1, F3 and F5).

Analysis of the Experimental Results in AUC. Analysis of AUC in Table 2 indicates the following. First, the performance of the proposed model CLVHUP is superior to the other methods. The average AUC of CLVHUP reaches 0.7599. Second, MLP, FM and CLVHUP are three of best performers and their performances are closed. Specifically, the average AUC of each model is larger than 0.6858. In contrast to the other models, they have an obvious advantage in performance. These comparisons demonstrate that our work is meaningful. Analyzing payment behaviors from the perspective of CLV is beneficial for improving performance and demonstrates that deeply mining information underlying CLVs in CLVHUP is effective.

Table 3. Experimental results in Recall F1, F3 and F5.

Games	Models	Metrics			
		Recall	F1	F3	F5
15	MLP	0.2968 ± 0.0787	0.2149 ± 0.0441	0.272 ± 0.0588	0.2864 ± 0.0695
	FM	0.3633 ± 0.0546	0.2459 ± 0.037	0.3306 ± 0.0472	0.3499 ± 0.0513
	CLVHUP	**0.4984 ± 0.045**	**0.2538 ± 0.0348**	**0.4173 ± 0.0427**	**0.4637 ± 0.044**
59	MLP	0.1985 ± 0.0516	0.1447 ± 0.0133	0.183 ± 0.0385	0.1921 ± 0.0461
	FM	0.0982 ± 0.0206	0.1136 ± 0.0172	0.1005 ± 0.0187	0.099 ± 0.0198
	CLVHUP	**0.2234 ± 0.0408**	**0.1614 ± 0.0117**	**0.2064 ± 0.0295**	**0.2164 ± 0.036**
98	MLP	0.4226 ± 0.0405	0.2015 ± 0.0203	0.3449 ± 0.0239	0.3886 ± 0.032
	FM	0.3314 ± 0.027	**0.2513 ± 0.0161**	0.3113 ± 0.0227	0.3234 ± 0.0252
	CLVHUP	**0.471 ± 0.0198**	0.248 ± 0.0209	**0.3986 ± 0.0174**	**0.4402 ± 0.0179**
108	MLP	0.2391 ± 0.0773	0.1445 ± 0.0426	0.2081 ± 0.062	0.2258 ± 0.07
	FM	0.2641 ± 0.0552	**0.241 ± 0.0601**	0.2581 ± 0.054	0.2617 ± 0.0545
	CLVHUP	**0.3161 ± 0.0619**	0.2405 ± 0.0571	**0.2946 ± 0.0534**	**0.3072 ± 0.0576**
111	MLP	0.1669 ± 0.0326	0.1873 ± 0.0254	0.1704 ± 0.031	0.1682 ± 0.032
	FM	0.1935 ± 0.042	0.1945 ± 0.0332	0.1935 ± 0.0401	0.1935 ± 0.0413
	CLVHUP	**0.4105 ± 0.0346**	**0.2274 ± 0.0209**	**0.3532 ± ± 0.0286**	**0.3863 ± 0.0318**
163	MLP	0.3181 ± 0.0196	0.3236 ± 0.0111	0.319 ± 0.0175	0.3184 ± 0.0188
	FM	0.3176 ± 0.0148	0.3281 ± 0.0131	0.3196 ± 0.0142	0.3183 ± 0.0145
	CLVHUP	**0.3444 ± 0.0198**	**0.3468 ± 0.0095**	**0.3448 ± 0.0175**	**0.3445 ± 0.0189**
Avg	MLP	0.2737	0.2027	0.2496	0.2632
	FM	0.2613	0.2291	0.2523	0.2576
	CLVHUP	**0.3773**	**0.2463**	**0.3358**	**0.3597**

Analysis of the Experimental Results in Recall, F1, F3 and F5. We compared the AUC models in Table 2 and find that the performance of MLP, FM and CLVHUP are closed. Next, we analyze the performances of MLP, FM and CLVHUP in the metrics of Recall, F1, F3 and F5. First, we analyze Recall, F3 and F5 of these models. CLVHUP is the best, the there exists a large difference between CLVHUP and the other models. Second, we analyze the F1 of these models. The performance of FM and CLVHUP are closed, but CLVHUP performs better in average. These comparisons show that CLVHUP has the best comprehensive performance in practice.

Analysis of the Influence of the Classification Threshold. Since CLVHUP has the highest AUC, we question whether we can improve the performance of CLVHUP by tuning the classification threshold. In Fig. 3, we analyze the influence of the threshold. We find that the threshold is different when CLVHUP gains the best performance in F1 and F5. In summary, CLVHUP achieves the best performance in F1 when the threshold is equal to 0.5 and achieves the best performance in F5 when the threshold is equal to 0.2.

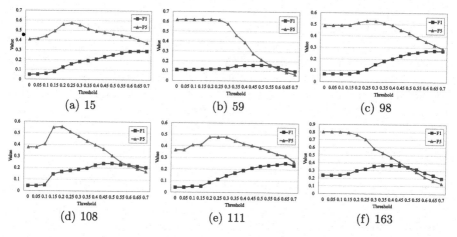

Fig. 3. Influence of threshold

6   Related Works

To the best of our knowledge, this is the first work to analyze payment behavior under the CLV dimension and to predict high-value users by utilizing machine learning techniques and exploring the information underlying CLV. The works most related to ours are those of Drachen et al. [2] and Sifa et al. [5]. Drachen et al. found many social features and then incorporated them with traditional models (random forest and XGBoost) to predict game customer lifetime value. However, they did not propose a novel and suitable model to address this challenge and ignored the information underlying the CLV. Sifa et al. first used classification to predict whether a purchase will occur, and second, used a regression model to estimate the number of purchases a user will make. They did not explore the information underlying the CLV. In e-commerce, Vanderveld et al. [6] and Chamberlain et al. [1] aimed to identify high-value users in e-commerce. Vanderveld et al. were the first to address the CLTV problem. They first split the original problem into several simpler subproblems and then solved these subproblems to address the original problem. Chamberlain et al. proposed a novel method for generating the embedding of customers, which addresses the issue of the ever-changing product catalog and obtained a significant improvement over an exhaustive set of handcrafted features. The most considerable difference between e-commerce and games is that the lifetime of users in games is much shorter and includes less data, which increases the difficulty in solving HU prediction (must identify high-value users when they are low level; otherwise, they will probably leave the game in the future).

7   Conclusions

In this paper, we first analyze payment behavior under the CLV dimension. Then, to predict HU, we propose a novel CLV-based model, CLVHUP, which captures the information underlying CLV. In a series of experiments, we verify the effectiveness of CLVHUP, which outperforms many techniques (LR, SVM, FM and MLP). In the future, we plan to experiment on more datasets and mine more and effective features from user data of games.

References

1. Chamberlain, B.P., Cardoso, Â., Liu, C.H.B., Pagliari, R., Deisenroth, M.P.: Customer lifetime value prediction using embeddings. In: Proceedings of the 23rd ACM SIGKDD International Conference on Knowledge Discovery and Data Mining, Halifax, NS, Canada, 13–17 August 2017, pp. 1753–1762. ACM (2017)
2. Drachen, A., et al.: To be or not to be... social: incorporating simple social features in mobile game customer lifetime value predictions. In: Proceedings of the Australasian Computer Science Week Multiconference, p. 40. ACM (2018)
3. Rendle, S.: Factorization machines. In: 2010 IEEE 10th International Conference on Data Mining (ICDM), pp. 995–1000. IEEE (2010)
4. Rendle, S., Gantner, Z., Freudenthaler, C., Schmidt-Thieme, L.: Fast context-aware recommendations with factorization machines. In: Proceedings of the 34th international ACM SIGIR Conference on Research and Development in Information Retrieval, pp. 635–644. ACM (2011)
5. Sifa, R., Hadiji, F., Runge, J., Drachen, A., Kersting, K., Bauckhage, C.: Predicting purchase decisions in mobile free-to-play games. In: Proceedings of the Eleventh AAAI Conference on Artificial Intelligence and Interactive Digital Entertainment, AIIDE 2015, 14–18 November 2015, University of California, Santa Cruz, CA, USA, pp. 79–85 (2015)
6. Vanderveld, A., Pandey, A., Han, A., Parekh, R.: An engagement-based customer lifetime value system for e-commerce. In: Proceedings of the 22nd ACM SIGKDD International Conference on Knowledge Discovery and Data Mining, pp. 293–302. ACM (2016)

Author Index

Printed in the United States
By Bookmasters